D1172017

NUCLEAR POWER

Radiation

JAMES A. MAHAFFEY, PH.D.

Facts On File
An Infobase Learning Company

For Jack Alexander Fletcher

RADIATION

Facts On File
An imprint of Infobase Learning
132 West 31st Street
New York NY 10001

Library of Congress Cataloging-in-Publication Data
Mahaffey, James A.
    Radiation / James A. Mahaffey.
       p. cm.
    Includes bibliographical references and index.
    ISBN 978-0-8160-7652-9
1. Radiation. I. Title.
    QC475.25.M34 2011
    539.2—dc22   2010049237

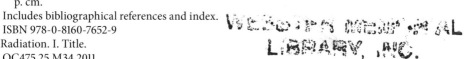

Facts On File books are available at special discounts when purchased in bulk quantities for businesses, associations, institutions, or sales promotions. Please call our Special Sales Department in New York at (212) 967-8800 or (800) 322-8755.

You can find Facts On File on the World Wide Web at http://www.infobaselearning.com

Text design by Annie O'Donnell
Composition by Adept Concept Solutions
Illustrations by Bobbi McCutcheon
Photo research by Suzanne M. Tibor
Cover printed by Bang Printing, Brainerd, Minn.
Book printed and bound by Bang Printing, Brainerd, Minn.
Date printed: October 2011
Printed in the United States of America

10 9 8 7 6 5 4 3 2 1

This book is printed on acid-free paper.

Contents

Preface

Nuclear Power is a multivolume set that explores the inner workings, history, science, global politics, future hopes, triumphs, and disasters of an industry that was, in a sense, born backward. Nuclear technology may be unique among the great technical achievements, in that its greatest moments of discovery and advancement were kept hidden from all except those most closely involved in the complex and sophisticated experimental work related to it. The public first became aware of nuclear energy at the end of World War II, when the United States brought the hostilities in the Pacific to an abrupt end by destroying two Japanese cities with atomic weapons. This was a practical demonstration of a newly developed source of intensely concentrated power. To have wiped out two cities with only two bombs was unique in human experience. The entire world was stunned by the implications, and the specter of nuclear annihilation has not entirely subsided in the 60 years since Hiroshima and Nagasaki.

The introduction of nuclear power was unusual in that it began with specialized explosives rather than small demonstrations of electrical-generating plants, for example. In any similar industry, this new, intriguing source of potential power would have been developed in academic and then industrial laboratories, first as a series of theories, then incremental experiments, graduating to small-scale demonstrations, and, finally, with financial support from some forward-looking industrial firms, an advantageous, alternate form of energy production having an established place in the industrial world. This was not the case for the nuclear industry. The relevant theories required too much effort in an area that was too risky for the usual industrial investment, and the full engagement and commitment of governments was necessary, with military implications for all developments. The future, which could be accurately predicted to involve nuclear power, arrived too soon, before humankind was convinced that renewable energy was needed. After many thousands of years of burning things as fuel, it was a hard habit to shake. Nuclear technology was never developed with public participation, and the atmosphere of secrecy and danger surrounding it eventually led to distrust and distortion. The nuclear power industry exists today, benefiting civilization with a respectable percentage of the total energy supply, despite the unusual lack of understanding and general knowledge among people who tap into it.

This set is designed to address the problems of public perception of nuclear power and to instill interest and arouse curiosity for this branch of technology. *The History of Nuclear Power*, the first volume in the set, explains how a full understanding of matter and energy developed as science emerged and developed. It was only logical that eventually an atomic theory of matter would emerge, and from that a nuclear theory of atoms would be elucidated. Once matter was understood, it was discovered that it could be destroyed and converted directly into energy. From thre it was a downhill struggle to capture the energy and direct it to useful purposes.

Nuclear Accidents and Disasters, the second book in the set, concerns the long period of lessons learned in the emergent nuclear industry. It was a new way of doing things, and a great deal of learning by accident analysis was inevitable. These lessons were expensive but well learned, and the body of knowledge gained now results in one of the safest industries on Earth. *Radiation*, the third volume in the set, covers radiation, its long-term and short-term effects, and the ways that humankind is affected by and protected from it. One of the great public concerns about nuclear power is the collateral effect of radiation, and full knowledge of this will be essential for living in a world powered by nuclear means.

Nuclear Fission Reactors, the fourth book in this set, gives a detailed examination of a typical nuclear power plant of the type that now provides 20 percent of the electrical energy in the United States. *Fusion*, the fifth book, covers nuclear fission, the power source of the universe. Fusion is often overlooked in discussions of nuclear power, but it has great potential as a long-term source of electrical energy. *The Future of Nuclear Power*, the final book in the set, surveys all that is possible in the world of nuclear technology, from spaceflights beyond the solar system to power systems that have the potential to light the Earth after the Sun has burned out.

At the Georgia Institute of Technology, I earned a bachelor of science degree in physics, a master of science, and a doctorate in nuclear engineering. I remained there for more than 30 years, gaining experience in scientific and engineering research in many fields of technology, including nuclear power. Sitting at the control console of a nuclear reactor, I have cold-started the fission process many times, run the reactor at power, and shut it down. Once, I stood atop a reactor core. I also stood on the bottom core plate of a reactor in construction, and on occasion I watched the eerie blue glow at the heart of a reactor running at full power. I did some time in a radiation suit, waved the Geiger counter probe, and spent many days and nights counting neutrons. As a student of nuclear technology, I bring a near-complete view of this, from theories to daily operation of a power plant. Notes and apparatus from my nuclear fusion research have been requested by and given to the National Museum of American History of the Smithsonian Institution. My friends, superiors, and competitors for research funds were people who served on the USS *Nautilus* nuclear submarine, those who assembled the early atomic bombs, and those who were there when nuclear power was born. I knew to listen to their tales.

The Nuclear Power set is written for those who are facing a growing world population with fewer resources and an increasingly fragile environment. A deep understanding of physics, mathematics, or the specialized vocabulary of nuclear technology is not necessary to read the books in their series and grasp what is going on in this important branch of science. It is hoped that you can understand the problems, meet the challenges, and be ready for the future with the information in these books. Each volume in the set includes an index, a chronology of important events, and a glossary of scientific terms. A list of books and Internet resources for further information provides the young reader with additional means to investigate every topic, as the study of nuclear technology expands to touch every aspect of the technical world.

Acknowledgments

I wish to thank Dr. Douglas E. Wrege and Dr. Don S. Harmer, from whom I learned much as a student at the Georgia Institute of Technology in the schools of physics and nuclear engineering. They were kind enough to read the rough manuscript of this work, checking for technical accuracy and readability. Their combined wealth of knowledge in nuclear physics was essential for polishing this book. The manuscript also received a thorough cleansing by Randy Brich, a most knowledgeable retired USDOE health physicist from South Dakota and currently the media point-of-contact for Powertech Uranium. Special thanks to Suzie Tibor for researching the photographs and to Bobbi McCutcheon for preparing fine line art.

Introduction

Radiation can be the "elephant in the room" when discussing nuclear power, or it can be the central topic of a much wider debate concerning general nuclear technology. The primary danger of nuclear science has always seemed to be radiation, the destructive by-product of *fission*. The act of fissioning a uranium or *plutonium* nucleus releases energy, and about 10 percent of this energy is in the form of intense, penetrating radiation. The entire measure of energy from fission can take thousands of years to fully materialize, and therein lies a problem. Long after the fission has occurred to produce power in a nuclear reactor, after the power plant has worn out and been torn down, and after the ground on which the power plant sat has been seeded in grass and returned to nature, a weak echo of the power production can still be felt in the remaining fission by-products. It is this lingering hint of danger that must be studied and understood for a complete survey of nuclear power and the technology that makes it possible.

Radiation, one volume in the Nuclear Power set, explains the nature of radiation in its many forms in the first chapter. Chapter 2 goes into detail concerning what is and what is not dangerous about radiation, explaining its many effects in matter, in both living and nonliving things. Radiation has always occurred in nature, bombarding us from birth, and this phenomenon is discussed in chapter 3. The many industrial uses of radiation, from smoke detectors to dental X-rays, are explained in chapter 4. The many techniques that are used to detect and measure this invisible phenomenon are covered in chapter 5. Practical measures of radiation protection, from elaborate shielding to simply avoiding it, are discussed in chapter 6, and an introduction to medical therapies for radiation poisoning in the last chapter gives assurance that acute radiation exposure can be treated. From this broad survey of radiation, its detection, its effects, and protection against it, a clearer understanding of the challenges of a nuclear power economy is possible.

Some interesting sidelights of the main discussion are covered in sidebars in each chapter, and photographs and diagrams provide an increased clarity of major points. Depictions of *radioactivity* intensity, dose, and rate of dose accumulation are a complicated mixture of units and subtle meanings. Although there is a firmly instituted system of measurement, the *Système international d'unités (SI),* these units have been largely ignored in the nuclear industry in the United States. There are presently three countries that have not officially accepted the SI system of scientific units: Liberia, Myanmar, and the United States. Most existing radiological instruments and even many new instruments are calibrated in the traditional *roentgen* system and to read a great body of scientific papers and texts requires knowledge of this now antiquated system. These radiation units are carefully unraveled in this volume, with explanations of the different types of measurement for each unit, always given in terms of both systems, the traditional and the SI.

Mathematical representations are kept to a minimum, but there is much new terminology necessary for these discussions. All technical terms, unique to this volume and as used in the other books in this series, are included in the glossary for quick reference. The chronology is a mixture of invention and discovery dates for radiation types and measurement techniques, as well as a record of important dates in the development of our understanding of radiation and its effects.

In these discussions of radiation, an *element* is a fundamental material, such as carbon, gold, or nitrogen. Each element has its own characteristic atomic *nucleus,* consisting of a set number of protons, which determine its chemical characteristics, but a variable number of *neutrons,* determining its radioactivity characteristics. Each element has several subspecies, or *nuclides,* each having the same number of protons but a characteristic number of neutrons. *Isotope* and *nuclide* are used in many discussions of nuclear topics, and their meanings are basically identical, but *isotope* is gradually becoming archaic. A single isotope can have two characteristic nuclides, as is the case with technetium-99. The nuclide technetium-99m is the metastable isomer of technetium-99. Both species of the element technetium have 56 neutrons in the nucleus, but both are radioactive in different ways, with different half-lives and different particle emissions.

A specific nuclide is designated by the element name with an appended whole number. The number is the *mass number* of the nuclide, or the sum of the numbers of protons and neutrons in the nucleus. The element carbon, for example, has 10 known isotopes or nuclides, from carbon-9 to carbon-18. Each carbon nuclide has six protons in the nucleus, but there may be anywhere from three to 12 neutrons in a carbon nucleus, with various outcomes. Most of the carbon on Earth, 98.89 percent, is carbon-12, with almost the entire rest of the carbon being carbon-13. These carbon nuclides are stable and nonradioactive, but all other nuclides of carbon, such as carbon-14, are unstable. The unstable nucleus in carbon-14 is likely to decay, or break down into another nucleus and another element. It is therefore considered radioactive.

1 Types and Sources of Radiation

Radiation is invisible. It has no taste, no smell, no texture, and makes no sound. For tens of thousands of years human beings were completely unaware of it, although everyone was being constantly bombarded with a wide range of radiation types from all directions. Radiation rained down from the sky, and it projected up, out of the ground and from nearby rock formations. It was and still is in the food, in the water, and in the air being breathed. Mankind was blissfully unaware of its presence in everyday life.

In 1864, James Clerk Maxwell (1831–79), a Scottish theoretical physicist and mathematician, predicted the existence of electromagnetic waves in a purely mathematical, nonexperimental exercise. This type of wave was thought to exist in theory, but it had never been observed or measured.

At the University of Karlsruhe, Germany, in 1887, Heinrich Hertz (1857–94) accidentally confirmed the existence of Maxwell's theoretical waves, finding that these electric waves behaved exactly as the mathematical equations predicted they would. Invisible and undetectable by ordinary senses, these artificially produced waves traveled with the speed of light and performed with optical characteristics. It became clear that they were of the same nature as light but of lesser energy and vibrational frequency. The phenomenon was eventually named radio waves.

In 1895, Wilhelm Roentgen (1845–1923), a German physicist at the University of Würzburg, made a further discovery of Maxwellian waves, but these invisible rays were of higher energy and frequency and capable of penetrating solid objects in ways that visible light could not. The new type of waves was produced artificially, using high-voltage electrical equipment, vaguely similar to Hertz's setup. As a temporary measure, Roentgen named his discovery *X-rays*. It was quickly found that in large quantities these higher-energy waves could be harmful to living things.

Hoping to improve on Roentgen's discovery, the Frenchman Henri Becquerel (1852–1908) accidentally discovered an even higher energy electromagnetic wave in 1896. These highly penetrating waves required no high-voltage electrical apparatus for production. They seemed to emanate without stimulus from an ordinary mineral, used as a dye in ceramics. The active ingredient in the mineral was *uranium*. With close and rigorous study of the phenomenon, a student of Becquerel, Marie Curie (1867–1934), would confirm the wave-emitting nature of this and other mineral components and give it a lasting name—radiation.

Once it was firmly established that radiation can be produced both artificially and by naturally occurring substances, increasing research in the early 20th century revealed more types of radiation occurring in a wider range of energies, and the effects of radiation on living and nonliving matter were measured. Advanced theories in the 1930s further clarified the nature and the sources of radiation, and the discovery of nuclear fission in 1939 increased the importance of this knowledge, as the release of nuclear energy came about soon after. The importance of radiation, particularly in a study of nuclear power, is due to the inherent danger of this by-product of energy release. For safety in a world of increasing nuclear power generation, knowledge of radiation is essential.

In this first chapter, the long list of radiation types is unraveled, including possible sources for each type and some basic specifications. Sidebars cover two rare offshoots of the taxonomic tree of radiation classification, the exotic particles and solar flares, which are a clear danger only to space travelers.

ULTRAVIOLET, GAMMA, AND X-RAYS

Before discussing the various types of radiation, it is important to explain the unit of measure used to express the energy of individual particles or rays. This unit is the electron volt (eV). The electron volt is not an SI unit, but it is most often used with SI multiplicative prefixes, particularly kilo (1,000), mega (1,000,000), giga (1,000,000,000), and occasionally tera (1,000,000,000,000). These prefixes are abbreviated to one letter, or *KeV*,

MeV, GeV, and *TeV.* Adding to the confusion, eV can also be used as a unit of mass, because it has been proven that mass and energy are equivalent. For the sake of convenience, the conversion factor between mass and energy, or the speed of light squared (c^2), is simply set to 1. In this case, the mass of an electron, for example, is 0.511 MeV. A positron, or antielectron, is the same mass. When the two meet, they annihilate each other completely, with both masses reverting to pure energy, or twice 0.511 MeV. The energy released in the destructive meeting of an electron and a positron is thus 1.022 MeV.

One electron volt equals 1.6×10^{-19} joules of energy, or the energy required to raise one free electron to a potential of one volt. To put this in perspective, the total energy released in the fission of one uranium-235 nucleus is, on average, 200 MeV, or 200 million electron volts. The nuclear *fusion* of one deuterium nucleus and one *tritium* nucleus to form one helium-4 nucleus releases, on average, 17.6 MeV. One snowflake hitting the concrete pavement releases about 4 MeV of heat. To ionize a hydrogen atom, or knock its single electron out of orbit, requires 13.6 eV. The combined energy of two protons accelerated to ramming speed in the Large Hadron Collider (LHC) near Geneva, Switzerland, is 14 TeV. The energy of the most energetic particle known to exist, the ultra-high-energy cosmic ray, is somewhere more than 10 million TeV.

A single, energetic moving particle can also be thought of as an oscillating wave. In fact, the concepts of particle and wave, in the very small world of atoms and subatomic particles, are interchangeable. On the subatomic scale, all rules or anti-rules of quantum mechanics apply, and this means that a particle can be a wave or a ray, and a wave or a ray can be a particle. If a particle is a wave, then it has a frequency and a wavelength. This vibrational characteristic is usually expressed as wavelength in nanometers, or billionths of a meter. The equation for energy to wavelength conversion is:

$$E = 1,240/\lambda,$$

where *E* is the particle or wave energy in electron volts (eV), and λ is the wavelength in nanometers (nm). The number 1,240 is *Planck's constant,* adjusted to work with eV and nanometer notations. As the wavelength becomes shorter, the energy of the ray or particle becomes greater. As the energy increases, so do the destructive and penetrating qualities of the radiation. Visible light with a wavelength of 500 nanometers, which is green, is stopped by a human hand. X-rays at a wavelength of one nano-meter go completely through a hand and are stopped only by the bones.

This floating designation of particle or ray makes it difficult to classify radiation types, yet there is one large category of radiation occurring on a continuous spectrum of electromagnetic rays. The most familiar form of electromagnetic radiation, other than visible light, is radio and television transmission. Every time a cell phone call is made, a sphere of radiation develops at the antenna of the telephone, usually at the top, traveling out-ward at the speed of light, 186,000 miles per second (3×10^8 m/s). At a rate of about 1×10^9 times per second, or 1 gigahertz, the radiation switches from an electrical charge caused by a changing magnetic field to a mag-netic field caused by a changing electrical charge. The radiation from a cell phone, or any other radio transmitter, is thus called *electromagnetic radiation*. The physical length of one complete wave of this electricity-to-magnetism transition, caused by a cell phone conversation, as it speeds through space is about 30 centimeters. Because all electromagnetic rays travel at the same speed in a given medium, the speed of light, the wave-length is always related to the frequency of vibration by the equation:

$$f = 3 \times 10^{15}/\lambda,$$

where *f* is the frequency in hertz, or cycles per second, and λ is the wave-length in nanometers divided into the approximate speed of light in a vacuum in nanometers per second. As a worldwide standard that predates the SI unit conventions of 1954, electromagnetic radiation wavelengths are always expressed in meters or prefixed fractions of meters.

At the lowest end of the electromagnetic spectrum, or the end with the lowest frequencies and lowest energy per particle, are radio waves. This is also the end with the longest wavelengths. Radio waves are commonly produced by inducing time-varying magnetic and electric fields in a metal antenna, using an alternating electrical current. The sub-spectrum of radio waves, used extensively for wireless communications and astronom-ical observations, ends in the spectrum somewhere above a frequency of 100 gigahertz and a wavelength of about one millimeter. The spectrum begins at three hertz, the lowest experimental frequency ever used in the *extremely low frequency (ELF),* band, extending from three to 3,000 hertz. ELF waves occur in nature at a frequency of 7.8 hertz, caused by lightning strikes, with a wavelength of the circumference of Earth, 24,819 miles

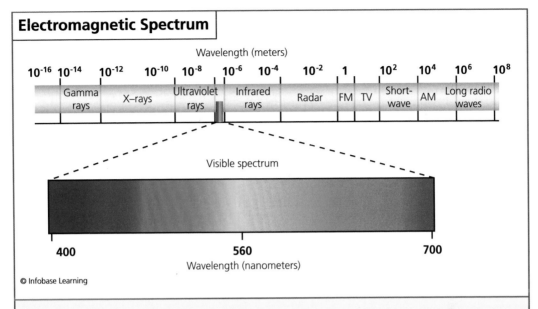

Electromagnetic Spectrum

Wavelength (meters)

10^{-16} 10^{-14} 10^{-12} 10^{-10} 10^{-8} 10^{-6} 10^{-4} 10^{-2} 1 10^{2} 10^{4} 10^{6} 10^{8}

| Gamma rays | X–rays | Ultraviolet rays | Infrared rays | Radar | FM | TV | Short-wave | AM | Long radio waves |

Visible spectrum

400   560   700

Wavelength (nanometers)

© Infobase Learning

The electromagnetic spectrum, from low-frequency radio waves to gamma rays—the visual light portion of the spectrum is shown in detail.

(4×10^7 m). The energy of these waves is so low that there are no known health effects from being exposed to them or any other radiation in the lower end of the radio spectrum. Practical applications of this extremely long radio wave are limited, as the length of an efficient antenna must be half the length of one cycle. The entire Earth, therefore, must be used as an antenna.

Using the Earth as a radiating medium means that the waves travel through the ground and through the salt water in the oceans. For this reason, a signal from an ELF transmitter in Clam Lake, Wisconsin, can be picked up anywhere in the world, even under water, and this unique feature of ELF waves was put to practical use for nuclear submarine communications. Unlike any other radio transmissions, the signals can be picked up by a completely submerged submarine, but the signal goes only one way. The submarine can receive, but it cannot transmit, and the information bandwidth at these frequencies limits digital communication to about 450 words per minute.

The *low frequency (LF)* radio band is higher up the electromagnetic spectrum, extending from 30 kilohertz to 300 kilohertz and is used for commercial broadcasting in Europe, Africa, the Middle East, and Mongolia. The wavelength range is 10 kilometers to one kilometer, and the necessary antennae are inconveniently long. In the Western Hemisphere, these low frequencies are used only for time signal stations and aircraft beacons.

Medium wave (MW) electromagnetic radiation has been used as the AM (amplitude modulation) radio broadcast band signal worldwide since the 1920s. In the United States, the MW broadcast band extends from 530 kilohertz to 1,700 kilohertz. The wavelength is about 300 meters. MW radio signals tend to follow the curvature of Earth, and this makes it ideal long-range radio service. A powerful broadcast-band transmitter of 50 kilowatts can transmit in a circle with a radius as long as 250 miles (400 km).

Above medium waves on the electromagnetic spectrum are *high frequency (HF)* radio waves. This band of frequencies, from three to 30 megahertz, is historically known as the shortwave band, because of the diminutive size of the waves. Shortwaves range from 100 to 10 meters long. Unique propagation properties of these waves have made them useful for very long-range overseas communications. A relatively weak shortwave signal of a few watts has sufficient energy to span oceans and hemispheres. These waves do not follow Earth's contour, as do medium waves, but they are refracted into a curvature-following path by the ionosphere, a region in the upper reaches of the atmosphere. The speed of electromagnetic radiation is slightly faster in the ionized gases of the ionosphere than in the lower, nonionized atmosphere, and this causes the short radio waves to be bent downward as they make the transition between the two different media.

Shortwave radiation transmission was developed by amateur radio enthusiasts trying to get away from the extreme static in the upper MW band as they attempted long-range communications at low power. In June and July 1923, Guglielmo Marconi (1874–1937), an Italian radio pioneer, successfully communicated with his yacht, *Electra,* in the Cape Verde Islands from a radio station in Cornwall, England, transmitting in the 97-meter band. By 1924, amateurs were routinely making transoceanic radio contacts at distances more than 6,000 miles (ca. 9,700 km), using radio signals of a few watts transmitted at 80, 40, and even 20 meters.

Still farther up the spectrum, radio radiation gains frequency, bandwidth capability, energy per particle, and general utility. The *very high frequency (VHF)* waves are in the spectral band from 30 to 300 megahertz. Wavelengths in the VHF band range from 10 meters down to one meter, and the VHF radiation carries both FM (frequency modulation) radio and television. FM radio is in the fairly narrow band from 88 to 108 megahertz, with VHF digital television skirting the radio signals both below and above the band. VHF radiation is not bent by the ionosphere, nor does it follow Earth's curvature. It travels only in a straight line, and any solid object is opaque to it. A VHF transmission is limited to the distance to the horizon, in what is called line-of-sight radio. Beyond the horizon, VHF radiation continues on in an unmodified direction, over the tops of radio antennas and out into space. The VHF band is used extensively, for everything from radio-controlled models to air traffic control.

The VHF waves are short enough to be concentrated into a collimated, parallel beam of radiation using parabolic reflectors, and it is possible to suffer a burn from close proximity to such an antenna while it is active. Most VHF radiation is purposefully generated, using electronic radio transmission equipment, but there is also a natural source of waves in this band. Celestial radio sources, from the Sun out into the Milky Way galaxy and beyond, constantly bombard Earth with VHF radiation.

Above VHF is the *ultrahigh frequency (UHF)* band, or UHF radio, in frequencies from 300 megahertz to three gigahertz. The wavelength is noticeably shorter in this radio band, from one meter down to 10 centimeters. This radio band is useful because of its very small wavelength and the correspondingly small antennae that are required to transmit and receive radiation in this frequency band. It is particularly applicable in short-range applications such as cell phones, WiFi wireless Internet connections, Bluetooth headsets, and garage door openers. The global positioning system (GPS) satellites transmit from geosynchronous orbits to Earth using UHF frequencies 1.57542 and 1.2276 gigahertz. High-definition television signals have moved from the VHF band up to the UHF band.

The practical end of radio participation in the electromagnetic spectrum occurs between three and 300 gigahertz, as the wavelength goes down to one millimeter. Beyond that frequency and wavelength, it seems impossible to make alternating current by strictly electronic means and excite a transmitting antenna. This radiation band is called microwave for its extremely short wavelengths. The microwave and UHF band designations tend to overlap in the low microwave and high UHF regions. A microwave oven, for example, actually uses radiation at 2.45 gigahertz, which could be classified as UHF, but is in the microwave S-band near the lower boundary of microwaves. Above microwave frequencies, in the submillimeter wavelength region, Earth's atmosphere is opaque, and the transmission of information by wireless means becomes impractical.

Microwaves are radiation of sufficient penetrating power and energy per particle that they have definite effects on living things. A two-second burst of 95 gigahertz microwaves aimed at a person will penetrate the skin to a depth of 1/64th of an inch (0.4 mm) and raise the temperature of the tissue to 130°F (54°C), causing a great deal of discomfort.

Beyond 300 gigahertz on the electromagnetic spectrum there are no radio waves, but in the region between microwaves and infrared light is *terahertz radiation (T-rays)*. T-rays occur in the band between 300 gigahertz and three terahertz. This corresponds to wavelengths between one millimeter, at the highest edge of the microwave band, and 100 micrometers. Terahertz radiation can pass through a wide variety of nonconducting materials, such as clothing, paper, cardboard, wood, masonry, plastic, and ceramics. Air is a strong absorber, limiting its usefulness for communication at a distance, but this radiation lacks enough energy to cause ionization, making it attractive for medical imaging. This nonionizing property means that, unlike X-rays or higher frequency radiation, terahertz rays do not damage tissues, cause burns, or disrupt DNA molecules. Many exotic methods of generating terahertz waves have been developed, such as backward wave oscillators, quantum cascade lasers, and solid state oscillators. The T-wave band remains an open subject of research.

Above terahertz rays on the spectrum are *infrared* rays. Although infrared rays are considered to be of the same sub-spectrum type as visible and *ultraviolet light,* they are beyond the ability of human eyes to perceive. The term *infrared* literally means "below red," with the color red being light of the lowest frequency or longest wavelength that humans can see. The infrared wavelength band is wide, running from 100 micrometers, the end of the terahertz band, to 750 nanometers, the threshold of visible red light. Frequencies are tens to hundreds of terahertz. A source of infrared radiation, visible light, and ultraviolet rays is an object raised to a temperature, with the frequency of the radiation being proportional to the temperature of the radiation emitter. A human being at normal body temperature emits infrared radiation at a wavelength around 10 micrometers. Direct sunlight contains the entire light spectrum, with 47 percent in the infrared band, 46 percent in the visible band, and 6 percent in the ultraviolet band.

Infrared radiation was discovered as an invisible component of sunlight in 1800 by William Herschel (1738–1822), an astronomer with the

Royal Society of London. The term *infrared* was not introduced until later in the 19th century. Applications of infrared radiation are many, including military night vision, astronomical observations, television remote controls, and remote temperature sensing. The sidewinder air-to-air missile, used to shoot down enemy aircraft, senses and tracks jet exhaust as a bright source of infrared rays.

Visible light is in the frequency range of petahertz, or 10^{15} hertz, and the wavelength begins at 750 nanometers, the end of the infrared band, and ends at 380 nanometers, or just beyond the color violet. Green light is around 500 nanometers in wavelength, and yellow light is around 600 nanometers. Visible light is in a very narrow portion of the entire electromagnetic radiation spectrum, which ranges from wavelengths of 1,000 meters to 0.001 nanometers, but it is the only radiation that humans can detect directly and is thus familiar. Although there was no need to discover light, an understanding of its nature has proceeded through many centuries. The English scientist Sir Isaac Newton (1643–1727) was one of the early researchers, publishing his *Hypothesis of Light,* describing light in terms of particles of matter, emitted in all directions, in 1675.

Visible light and all electromagnetic radiation are now thought of as both waves and hard particles, having measurable momentum. This radiation, in fact, exerts pressure. Light pressure from the Sun, for example, can affect asteroids, affecting speeds and trajectories. The light particle is the *photon.*

Above visible light, beyond the ability to detect colors of light, is the ultraviolet radiation band in the wavelength range from 380 nanometers down to 10 nanometers. Ultraviolet light is of sufficient energy to collide with an atomic-bound electron and knock it out of orbit, thus ionizing an atom, or imparting in it a net positive charge. This effect was discovered by the American physicist Arthur H. Compton (1892–1962) and earned him a Nobel Prize in 1927. The process of ionization is capable of damaging living tissues, and everything above visible light is in the class of ionizing radiation. Radiation in this class has significant individual particle energies, and ultraviolet photons range from three electron volts at the low-frequency end up to 124 electron volts.

Ultraviolet light causes sunburn, which is a characteristic skin burn of ionizing radiation. This feature of ionizing radiation is also commonly used for medical sterilization, disinfecting drinking water, and air purification. The ultraviolet band of radiation was discovered in 1801 by Johann Wilhelm Ritter (1776–1819), a German chemist and physicist, as an invisible component of sunlight beyond the violet end of the visible spectrum.

Moving still farther up on the electromagnetic radiation spectrum are X-rays, in the frequency range of 30 petahertz at the extreme end of the ultraviolet band to 30 exahertz, or 30×10^{18} hertz. Particle energies start at 120 electron volts and go up to 120 kilo electron volts, or 120,000 electron volts. Wavelengths are in the range of 10 down to 0.01 nanometers. X-rays are ionizing radiation with great penetration properties, and they are commonly produced using high-voltage vacuum tubes for use in medical and dental imaging. X-rays emitted from a nearby point source produce a strikingly clear image of bones and internal structures in the human body when converted to visible light by fluorescence on a flat screen or by exposure of a flat piece of photographic film.

Upon formal discovery in 1895, X-rays were immediately applied to medical science and in a short time were found to be useful in cancer treatments. It was discovered that while X-rays could permanently damage human tissues, cancerous tissue was more susceptible than normal tissues. By careful control of the total X-ray dosage, it was possible to reduce some types of cancerous tumors. Luggage scanners at airports use X-rays to examine the contents of closed cases and bags, and manufacturing industries use X-rays to examine welds for hidden flaws. In 1953, Rosalind Franklin (1920–58) at Kings College London used X-ray crystallography, recording the scatterings of X-rays by fixed nuclei in a molecule, to identify the structure of deoxyribonucleic acid (DNA).

The energy of X-rays, particularly near the upper spectral region, is significant. X-ray photon pressure, similar to visible light pressure but more intense, is used to hyper-compress deuterium and tritium in a thermonuclear weapon to cause explosive hydrogen fusion. In this case, X-rays are produced by a pulse of nuclear fission in a pluto-

Gamma

© Infobase Learning

The emission of a gamma (γ) ray from an atomic nucleus; the structure of the nucleus has shifted, and the movement of electrical charge under strong nuclear force has generated a powerful electromagnetic ray.

nium-based explosive used as a trigger for the fusion. As a nuclear fission explosion develops, in a progression of nanoseconds or billionths of a second, X-rays are the first radiation to escape the explosion. The X-ray wavefront is formidable, squeezing hydrogen-isotope nuclei together and producing fusion.

Last and highest on the scale of the electromagnetic spectrum are *gamma rays*. The frequency of these waves is on the order of tens of exahertz, or 10^{19} hertz, and the wavelength scale comes to an end at a little less than a picometer, or 10^{-12} meters. An alternate expression is 0.001 nanometers. Energies are more than 100 kilo electron volts, or 100,000 electron volts, reaching 10 mega electron volts. This value of 10,000,000 electron volts is often reduced to simply 10 MeV.

Gamma rays are simply a higher-energy form of X-rays, just as X-rays are a higher-energy form of ultraviolet rays, and so on down the spectrum, but gamma rays are produced by an entirely different source, and this gives them the unique notation. While X-rays are produced by energy transitions in the electron orbits of atoms or completely outside the atom, gamma rays are produced by energy transitions of protons and neutrons within the atomic nucleus. The potential for high-energy gamma rays is much greater, as the energy levels of the protons in a nucleus are generally much higher than electrons in atomic orbits, yet the highest energy, or hard, X-rays can actually be of greater energy than the weakest gamma rays. The distinction between gamma and X-rays can therefore be blurred and dependent entirely on the source of the radiation.

The following are ways in which a proton or neutron energy transition can occur in an atomic nucleus:

※ **Classical *Radioactive Decay***
Many atomic species of matter, or nuclides, are naturally capable of decaying to a different nuclide, always transforming into a different element. For example, the nuclide plutonium-238, which is one of 15 possible nuclides of the element plutonium, alpha decays into uranium-234. The probability of such a decay is expressed as the *half-life* of plutonium-238, which is 87.8 years. In 87.8 years, half a block of plutonium-238 will have changed into uranium-234. The process of an alpha decay severely changes the structure of the atomic nucleus. In rearranging itself from a torn-apart plutonium-238 to a stabilized uranium-234 nucleus, the nuclear particles settle into new,

lower energy states, emitting gamma rays at 0.04349, 0.091987, and 1.085 MeV. These gamma ray energies are unique to this decay of plutonium-238 and may be measured and used to identify this nuclide in an unknown sample. A beta decay example is the transmutation of the nitrogen-18 nuclide into oxygen-18. With a short half-life of 0.63 seconds, nitrogen-18 undergoes beta minus decay of the nucleus. In readjusting to the nuclide oxygen-18, two nuclear protons slip to lower energy states and emit gamma rays at 1.98 and 1.65 MeV.

❋ Nuclear Fission

Certain fissile nuclides, such as *uranium-235* and *plutonium-239,* can be induced to fission, or break in two, by incoming neutrons. The process involves a radical change of nuclear organization, as instead of transmuting to another nuclide of another element, the original nucleus becomes two distinct nuclides of two elements. One nucleus becomes two nuclei, both usually undergoing complete restructuring. For example, uranium-235 can fission into cesium-137 and rubidium-96. This reaction releases 193 MeV of energy, of which 4 percent is in the form of gamma rays.

❋ Spontaneous Fission

Closely related to neutron-induced nuclear fission, commonly used in nuclear reactors to produce power, is spontaneous fission, for which no external stimulus is necessary. Certain nuclide species of uranium, plutonium, and californium are capable of bursting roughly in half, emitting neutrons and gamma rays. In various nuclides of plutonium and uranium, this reaction is rare. For uranium-238, for example, the spontaneous fission half-life is 4.47 billion years. Californium-252, however, is extremely likely to undergo spontaneous fission, with a half-life of only 2.64 years. One gram of californium-252 emits 2.3 trillion neutrons per second from spontaneous fission. A small speck of californium-252 would be a fine, solid-state source of neutrons, but unfortunately it is a man-made element and one of the most expensive substances on Earth. The characteristic gamma rays from californium fission have energies 0.043, 0.100, and 0.160 MeV.

✳ **Neutron Capture**

The capture of a stray neutron by an atomic nucleus, or neutron activation, causes radioactive instability in the nucleus. The newly formed nuclide, now heavier by one neutron, emits a gamma ray as the protons jockey for a more stable condition. An example is gold-197, or gold in its natural, mined form. If irradiated by free neutrons, gold-197 can readily absorb one and become gold-198. Seeking stability, the gold-198 ejects a gamma ray at 0.412 MeV. This phenomenon has been exploited for detecting the chemical composition of materials remotely. It has, for example, been used to explore for gold. The ground is subjected to neutron radiation from a neutron source, and the irradiated earth is then monitored for gamma rays of the characteristic 0.412 MeV energy. The dirt does not have to be dug up or in any way chemically treated for this analysis. If 0.412 MeV gamma rays are detected, then the irradiated dirt contains gold.

✳ **Nucleosynthesis**

Nucleosynthesis is the process of creating new atomic nuclei by fusion of two atomic nuclei. The result of two light nuclei fusing together is always a larger, heavier nucleus, and the transaction often leads to a gamma ray emission. This forming of elements from simpler components first occurred within a few minutes of creation, or the *big bang,* when the primordial plasma of this explosive event began to cool down below 2 trillion degrees, and small amounts of helium, lithium, and beryllium were formed. Large-scale nucleosynthesis did not begin until the largely proton, or hydrogen, matter in the universe condensed into stars, where the conditions for nuclear fusion, high pressure and temperature, were optimized. In a small- to medium-sized star, such as the Sun, hydrogen-1 fuses with hydrogen-2, or *deuterium,* to form helium-3, and this transaction results in an emitted gamma ray. Although it is small for a star, the Sun is 865,000 miles (1,392,000 km) in diameter, with a core temperature of 28,300,000° Fahrenheit (15,700,000° Kelvin). Heavier elements are made in heavier stars, using the *CNO cycle,* or the process that makes carbon-12, carbon-13, nitrogen-14, helium-4, and oxygen-15 out of hydrogen-1. At three points in the cycle, in which hydrogen fuses with carbon-12 to make nitrogen 13, with carbon-13 to make nitrogen-14, and with nitrogen-14 to make oxygen-15, gamma rays are ejected as the new nuclei seek stability. Heavier elements are assembled in heavier stars, using complex processes, and the heaviest elements are built in supernovae. A *supernova* is an explosion of a star, lasting no more than a few seconds, in which elements heavier than iron are produced. All of these nucleosynthesis processes involve gamma ray production.

Isomeric Transition

An atomic nuclide can be in an excited *meta state,* without a tendency to undergo nuclear decay. In this case, the nuclear components, protons and neutrons, simply rearrange into a more stable configuration and release gamma rays due to the change in energy levels of the protons. A well-known example of an isomeric transition occurs in technetium-99m. The *m* indicates a *metastable isomer* form of the nuclide technetium-99. The metastable form has a half-life of 6.02 hours and decays to non-metastable technetium-99, emitting gamma rays at 0.00217 and 0.1476 MeV. Technetium-99m is used extensively for medical imaging, being directly injected into the human bloodstream. The gamma rays from the decay of the metastable state are then detected by an imaging system that can map out the positions of technetium-99m atoms in the bloodstream, thus indicating where and to what extent blood is circulating in the vascular system. Technetium is a man-made element, available from only one source in the Western Hemisphere, the NRU nuclear reactor at Chalk River, Canada.

There are other, rare forms of gamma ray production, including internal conversion, double beta decay, and double electron capture. As producers of electromagnetic radiation, these modes of radioactive decay are not only no threat to living things, they are very difficult to detect.

CHARGED PARTICLES

The common types of radiation known to originate as bits of matter having a fundamental electrical charge are alpha and beta particles. Although either type can be thought of and designated as either ray or particle, the alpha, being large and consisting of an entire helium-4 nucleus, is most often referred to as a particle. A *beta ray* is either an electron or an antielectron, referred to as a positron. In a formal setting, the two radiations are designated β^- and β^+. The positron has many characteristics of an electron, including the mass, but it is positively charged, whereas an electron is negatively charged.

In the process of alpha decay, *alpha particles* are emitted by certain heavy elements such as uranium, thorium, actinium, plutonium, or radium. The loss of a large piece of the nucleus leaves it unstable, bumps the mass number down by four, and transmutes the element down to a lighter nucleus. In readjusting the nucleus to a more stable condition, at least one gamma ray is emitted. The resulting alpha radiation is considered extremely dangerous, even though it has very little penetrating power. An alpha particle can be stopped by a single thickness of printer paper. Although it is easily stopped by collision with an intervening atomic nucleus, an alpha particle typically has a great deal of energy, and the ionizing and heating effects of being hit by an alpha are unusually intense.

Beta rays are high-energy electrons or positrons emitted by certain radioactive nuclides. The beta ray electron, or beta minus particle, is the result of a neutron in a nucleus decaying into a proton. The positive charge of the resulting proton and the negative charge on the ejected electron add together to zero charge, canceling out, and resulting in the lack of charge on the original

Alpha

An entire helium nucleus breaks free of a heavy atomic nucleus, leaving it lighter by four units—the free helium nucleus (α) is an alpha particle.

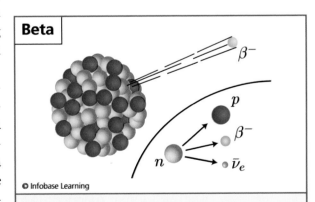

Beta

A beta minus, or $\beta-$, decay; a neutron breaks down into a proton, electron, and neutrino. Nothing is gained or lost in this transaction, as charge, mass, and energy are all conserved.

neutron. Similarly, a beta plus particle, or positron, is the result of a proton in a nucleus decaying into a neutron plus a positron, and the plus charge on the positron corresponds to the positive charge on the original proton. Charge, mass energy, and all aspects of the original nuclear particles are conserved in the beta ray transaction, with nothing lost or wasted.

In a beta minus decay, the nuclide retains its mass number but moves up one in atomic number. For example, carbon-14, with a 5,730-year half-life, decays upward on the periodic table to nitrogen-14 and releases an electron. The nitrogen-14 nucleus has essentially the same mass as the carbon-14 did, but it has seven protons in the nucleus instead of the six protons in the carbon nucleus, and its chemical characteristics are completely different. Nitrogen is a gas, and carbon is a solid, being a diamond or graphite in crystalline form. The carbon-14 beta ray has an energy of 0.156 MeV. Although the beta ray has more penetrating potential than an alpha particle, it is stopped completely by a thin aluminum plate.

A nuclide moves down one atomic number due to beta plus decay, which is a less common reaction. An example of a beta plus reaction is the decay of sodium-22, with a 2.601-year half-life. The emitted positron has an energy of 0.545 or 0.182 MeV, and a gamma ray of 1.275 MeV accompanies the decay. The result, or *daughter product,* of the decay is neon-22, a stable, nonradioactive inert gas.

NEUTRONS

The neutron is a constituent of the atomic nucleus and is present in every nuclide except hydrogen-1. As a building block of matter, it affects human beings only indirectly. As a dilution for the protons in a nucleus, too many or too few neutrons cause instability and a probability for radiation release of all types. As a free particle, far away from the nucleus, the neutron is radiation, with its own peculiar characteristics and interactions with intervening matter.

Free neutrons are produced by nuclear fusion or nuclear fission. The fusion of hydrogen-2, or deuterium, and hydrogen-3, or tritium, produces one helium-4 nucleus with an energy of 3.5 MeV and a free neutron with an energy of 14.1 MeV. The spontaneous fission of californium-252 produces an average of three neutrons, with a combined energy more than 6 MeV.

Neutrons are unique among free particles in that they have no charge and therefore are not affected by the negative electrical charge surrounding an atom or the positive charge surrounding a nucleus. These electrostatic charges are a serious detriment to moving alpha or beta particles and are known to stop them dead in a short distance. Neutrons, however, travel unaffected through the thickest, densest shielding material. A block of lead, which can stop the most energetic gamma ray, means nothing to a neutron. It just sails through it.

Neutrons do, however, interact with matter, even though the interaction has nothing to do with the strong electrostatic forces. A neutron is not deflected by the electrostatic repulsion or attraction when aimed at an atomic nucleus, and the flying particle will collide with a nucleus. Depending on the energy, or speed, of a flying neutron and the mass of the target it hits, it can seriously affect what it runs into. If the nucleus is lightweight, such as the single proton of hydrogen, then the neutron will make a perfect, billiard ball–style collision, exchanging momentum with the proton. If the proton is standing still when the accident occurs, it will go flying away with the energy of the neutron, and the neutron will be stopped cold. In this way, neutrons transfer their traveling energy to anything they hit, with the most efficient exchange being with something with similar mass, the hydrogen nucleus. Neutrons tend to bounce around in a solid mass, not traveling in a straight line but caroming off nuclei in the way, gradually losing their speed and slowing down to match the speed of the atoms that make up the material through which they are traveling. Neutrons can make it through heavy lead shielding, but not in a straight line. The trajectory is more of a drunkard's walk—a random series of direction and speed changes.

The limit to which a neutron's speed can be lost, from its initial speed of around one MeV, is the speed of target nuclei in the material. The material can be anything from air to steel, but the terminal speed is always the same. The neutron matches speed with the surrounding target material. All particles in all material move with thermal motion, at a speed determined solely by the temperature of the material. The average speed of a particle in a material is a definition of the temperature of that material. The faster the particles are moving, the hotter a material is. A *thermal neutron* has an energy of about 0.025 electron volts at room temperature and mixes in and moves with air particles. Thermalized neutrons are just like any other gas mixed in the air. Neutrons blow with the wind, rise in thermal columns, and can be breathed in and out of the lungs. Neutrons cannot, however, be confined under pressure in a steel gas cylinder, because they have no electrostatic force to hold them in.

A free neutron, no longer under the influence of an atomic nucleus, becomes a radioactive unit. It beta decays with a half-life of 10.6 minutes, emitting an electron of 0.782 MeV energy. The product of the radioactive decay is a stable proton, or a hydrogen-1 atom, ionized.

Neutrons affect matter in two ways. They impart heat to it, by simple collision, or they are captured by it, becoming part of the atomic nucleus of the target. If the target is fissile, as uranium-235 or plutonium-239, then there is a strong probability that the nucleus will break in half, releasing more energy than is justifiable by the total energy of the invading neutron. Capture by a non-fissile target, such as rock-stable mercury-202, makes mercury-203, which is highly unstable. Mercury-203 beta decays with a half-life of 46.60 days, ejecting a 0.212 MeV electron and a 0.2792 MeV gamma ray. The probability of neutron capture by a target is highly probabilistic and is dependent on the incoming energy of the neutron. As a rule, slow-moving neutrons are more likely to be captured, and the individual neutron-capture probability of a given nuclide at a given neutron speed is expressed in *barns*. The capture probability of a thermal neutron by mercury-202 is five barns, which is quite huge. The more typical neutron capture probability of nitrogen-14, of which 70 percent of air is composed, is 0.08 barns.

Neutron radiation can thus cause a secondary radiation effect, in which ordinary material exposed to the otherwise harmless, neutral-charged particles becomes radioactive, with all the potential dangers of radioactivity. A nearby blast of fast one-MeV neutrons causes internal heating, much the same way a microwave oven heats food without touching it.

GEOLOGICAL RADIATION SOURCES

Small concentrations of radioactive nuclides are present in all dirt and rocks on or near the surface of Earth. Radiation constantly bombards living things at a low level, coming from naturally occurring substances in the ground, water, air, and vegetation. Worldwide, the average natural background radiation dose for a human being is about 240 millirems (2.4 mSv) per year. In the United States, the average background is higher, at about 360 millirems (3.6 mSv) per year, due to surface deposits of uranium. About 0.5 millirems (0.005 mSv) of that radiation load is due to nuclear weapons testing, nuclear power accidents, and nuclear industry operations over the past 60 years. Medical tests, such as diagnostic X-rays, account for an average of 4.0 to 100.0 millirems (0.04 to 1.0 mSv) per year of background radiation exposure.

The largest component of this terrestrial background radiation comes from uranium-238 and *thorium-232,* which have been geologically present since Earth formed, about 4.5 billion years ago. Uranium-238 has a half-life of 4.5 billion years, which means that half of it has decayed away. Thorium-232, with a longer half-life of 14 billion years, has maintained a higher percentage of its initial geological presence. Both substances decay into chains of many intermediate radioactive nuclides, and the ends of both decay schemes are stable lead isotopes. The most notable radioactive daughter products of these decays are radium-224 and radon-220 from thorium and radium-226 and radon-222 from uranium. These radioactive nuclides are constantly present and bombarding human beings with radiation. The radium-226 nuclide from uranium, for example, emits alpha particles at 4.78 and 4.60 MeV and gamma rays at 0.19 and 0.26 MeV, with a half-life of 1,600 years.

Such radioactive sources in the geological formations of Earth are largely *self-shielding,* meaning that the heavy uranium nucleus can stop radiation as well as emit radiation. Alpha particles are available to the greater world only at the surface of minerals containing uranium or thorium, as alpha particles are stopped very quickly in solids. The gamma rays, while energetic with penetration power, also escape mainly from the surface and a very short distance under the surface. Minerals and even water are sufficiently dense to act as a shield against gamma rays. Beta rays from other members of the *decay chain,* such as bismuth-214 in the uranium decay, are far weaker than the gamma rays and are also emitting only from the surface and not the solid body of a radioactive rock.

The potential of radiation exposure from radon-222, however, is an entirely different matter. Radon is an inert gas unable to combine with or be captured by another substance. Radon-222 comes into being spontaneously, as the decay product of radium-226 in the decay scheme of uranium-238. The newly formed radon atom will make its way to the geological surface eventually unless it decays into solid (metallic) polonium-218 on the way, and it will leak into the atmosphere. Radon-222 is a particularly harmful natural substance, as it can be inhaled. It has a half-life of 3.82 days, and if it is in a lung when it decays, the delicate tissue of the lung submits to a powerful six MeV alpha particle. All of the energy of the particle is absorbed, and its full damage potential is realized. Moreover, the daughter product of the decay is polonium-218, a solid which

Pitchblende, the purest form of uranium ore *(DEA/PHOTO 1/Getty Images)*

does not readily exhale, and it has a 3.1-minute half-life with its own six MeV alpha particle. There are at least six more decays from that one radon atom, each emitting its own radiation until it becomes stable lead. According to the National Cancer Institute, the inhalation of natural radon is the second largest cause of lung cancer, behind cigarette smoking.

Natural uranium and thorium occur commonly in granite, which forms a major part of Earth's crust. Certain types of granite contain from 10 to 20 parts per million uranium, while sedimentary rocks, such as limestone, contain one to five parts per million uranium. If a house is built over granite bedrock and well insulated against cold, then radioactive radon gas can seep through the basement floor and be concentrated by the airtight nature of good insulation. A drafty dwelling, with more natural airflow, is much less likely to harbor radon. Radon also commonly collects in wells drilled through bedrock, as the radon from the uranium in the rock concentrates at the bottom where air is not moving and diluting it.

There are six primary uranium minerals in Earth's crust and nine secondary minerals containing detectable concentrations of uranium. The highest-grade uranium ore is uraninite, or pitchblende, occurring in the Athabasca Basin region of Canada. This black mineral is uranium oxide, which is known to occur also as the black specks in some types of granite. In the Oklo uranium mines in Gabon, Africa, the pitchblende is of sufficient purity to have achieved critical masses of uranium in Precambrian times, 5 million years ago. At that time, the uranium-235 content of the uranium-nuclide mixture was 3 percent, and this was sufficient fissile uranium to become a naturally occurring nuclear reactor when the ore was contaminated with groundwater. The uranium-235 content of natural uranium worldwide has dropped now to 0.7202 percent.

The only other naturally occurring geological radiation source is potassium-40. With a half-life of 1.28 billion years, only about 8 percent of the original potassium-40 in Earth's crust is still here. It emits a 1.31 MeV beta ray and a 1.46 MeV gamma ray. Potassium is essential to animal life, providing osmotic balance between cells and the interstitial fluid that separates them. We cannot live without potassium, and 0.012 percent of it is radioactive potassium-40. There is a detectable radiation load, for example, from eating a banana, which contains potassium. The nuclear "reference banana" contains 0.01 millirem (5.28×10^{-4} μCi) of radiation. A truckload of bananas contains sufficient radiation to set off nuclear weapon detectors at border-crossing checkpoints. The banana obtains potassium from the ground in which it grows.

Other known sources of potassium-40 in food are white potatoes at 3.40 microcuries per kilogram, sweet potatoes at 4.45 microcuries per kilogram, lima beans at 4.64 microcuries per kilogram, and spinach at 6.60 microcuries per kilogram. Technically, it is illegal for a person without a federal radioactive material license to possess a sample of more than 10 microcuries of potassium-40, so this could make the ownership of 4.4 pounds (2 kg) of spinach problematic.

Water, tumbling over rocks in a stream, slowly picks up all manner of soluble elements and compounds from the earth and eventually deposits them in the ocean. The U.S. Environmental Protection Agency (EPA) has maximum allowable levels of radioactive nuclides in drinking water. The combined concentrations of radium-226 and radium-228 cannot exceed a radiation loading of 19 picocuries per gallon (5 pCi/l). Uranium cannot be present in concentrations greater than 30 parts per billion (30 μg/l). A proposed standard for radon in water is 1,100 picocuries per gallon (300 pCi/l). Most areas of the United States have nothing close to this concen-

EXOTIC FORMS OF RADIATION

A survey of electromagnetic radiations, common charged particles, and neutrons would seem to cover the subject of possible radiation types, but actually it barely scratches the surface. When a nuclide undergoes radioactive disintegration, a cascade of exotic radiation and particle types split off and fly into space. Most types have no effect on human beings or the physical world and are of interest only to those who study and monitor the Standard Model of universal matter, to which these radiations mean a great deal. There are a couple of odd radiation particles that are difficult to ignore.

There is, for example, the neutrino, first theorized to exist in 1930 by Wolfgang Pauli (1900–58), an Austrian theoretical physicist at ETH Zürich, Eidgenössische Technische Hochschule Zürich (Swiss Federal Institute of Technology Zurich). The existence of a neutrino was necessary for beta decay, to account for all the loose components occurring as a result of this type of disintegration. In the beta decay of carbon-14, for example, a neutron becomes a proton in the nucleus, and a high-energy electron is kicked out of the nucleus. This transaction does not account for a net loss of energy and momentum, and that must have been carried away by another particle. The particle was named the neutrino, because it, of course, has no electrical charge. The positive charge of the proton plus the negative charge of the resulting electron add up to zero, the charge of the original neutron.

tration of radioactive nuclides in drinking water. However, certain western states, such as Utah and Colorado, where uranium has been mined, have more concerns about concentrations.

COSMIC RAYS

The only other radioactive nuclide, or *radioisotope,* naturally occurring in the environment is carbon-14. Carbon-14 emits a beta ray of 0.156 MeV with a half-life of 5,730 years.

Because terrestrial life is carbon based, every living thing contains a small amount of carbon-14. The concentration is extremely slight. Almost 99 percent of the world's carbon is carbon-12, 1 percent is carbon-13, and about one part per trillion is carbon-14. Anything that breathes or eats has a constant inflow and outflow balance of carbon, and the low saturation

The existence of the neutrino was confirmed in 1956 at the Savannah River National Laboratory in South Carolina by Clyde Cowan (1919–74) and Frederick Reines (1918–98). Having no electrical charge and traveling at near the speed of light, neutrinos have almost zero interaction with matter and can pass through the middle of an entire planet without disturbing anything. They are notoriously difficult to detect. Neutrinos are produced in great quantities by the fusion reactions in the Sun and stars and by nuclear fission in reactors on Earth. More than 50 trillion neutrinos pass through the human body every second.

There are actually three distinct types of neutrino: the electron neutrino, muon neutrino, and tau neutrino, each of which has an antimatter equivalent, or antineutrino. The muon, for which the second neutrino type is named, is another interesting radiation particle. The muon is an extremely heavy, negatively charged particle, having a mass of 105.7 MeV, or 200 times the mass of an electron. Outside the atomic nucleus, it has a half-life of 2.2 microseconds. The muon is so heavy, in fact, it cannot be produced by any Earth-bound nuclear disintegration event, including fission in a nuclear reactor or even in a nuclear weapon detonation. Muons come to us from outer space, as the decay products of extremely fast-moving protons as they crash into the atmosphere. The particle decays into an electron and two types of neutrino, an antielectron neutrino and a muon neutrino. There is naturally an antimuon, with a mirror-image decay scheme.

The muon was first detected by Carl Anderson (1905–91) at Caltech in 1936, during a study of cosmic rays.

level of carbon-14 in an organism is maintained until it dies, at which time the carbon content at death is fixed. The carbon-14 in a nonliving organism decays away with its 5,730 year half-life, and the decaying carbon-14 is never replaced. Therefore, the relative level of carbon-14 in a nonliving organism or organism component, such as a piece of wood, fossil fuel, or a human bone, can be carefully measured, and the level of beta radioactivity from the remaining concentration of carbon-14 can be used to date the organic material. If the wooden beam in an ancient building shows a beta activity of half what a new wooden beam shows, then that piece of wood is 5,730 years old. The older an organic sample is, the lower its carbon-14 signature, as the carbon-14 beta source becomes weaker with age.

If the world is 4.5 billion years old, then there should be no carbon-14 left from the original material from which this planet was formed. It should have been essentially gone by about 50,000 years of age. However,

A cosmic ray shower in Earth's atmosphere, caused by the impact of a single, high-speed particle from outer space *(Mark Garlick/Photo Researchers, Inc.)*

carbon-14 is constantly made and replenished in the atmosphere by the action of energetic particles from outer space that are impinging on the atmosphere. This class of radiation is called *cosmic rays*.

Soon after natural radioactivity was discovered by Henri Becquerel (1852–1908), background radiation was discovered, and it was assumed that it was the result of radioactive minerals in the ground. This was only partially correct, and in 1912 Victor Hess (1883–1964), an Austrian-American physicist, went aloft in a balloon carrying radiation detection instruments to study background radiation. If the radiation were only due to terrestrial sources, then the level should go down as altitude increased. It does not. The radiation level increases, indicating that there is another source outside the atmosphere. Hess was awarded the Nobel Prize in physics in 1936 for his discovery, and his newfound source of radiation was eventually named cosmic rays.

Cosmic rays bombard the Earth constantly from a variety of sources. The origins range from the Sun to unknown sources at the farthest range of the visible universe, and energies of individual particles can be extremely high, more than 10^{20} electron volts, or 100 eV. About 90 percent of all cosmic rays are high-speed protons, or hydrogen nuclei; 9 percent are alpha particles, or helium nuclei; and 1 percent are electrons. These are all charged particles and as such are heavily influenced by the geomagnetic field of the Earth, the same force that influences a magnetic compass. The greatest portion of these incoming rays is deflected away from the atmosphere. Without the influence of this magnetic field, the surface of the Earth would be sterilized of all life from the heavy flux of energetic radiation.

Cosmic rays that make it through the magnetic field hit the top of the atmosphere at full speed and typically crash into the nucleus of an oxygen atom. The result is destruction of the oxygen nucleus, with a spray of subatomic debris showering down through the air. The number of sub-particles created by a single cosmic ray hit can number in the billions. In the simplest process, a high-speed proton from a rotating neutron star, a supernova, or a black hole somewhere in the galaxy hits an oxygen-16 nucleus and tears out a neutron, leaving oxygen-15. Also originating in the collision are *pions,* having a positive, negative, or neutral charge, which proceed to decay into various forms of muons and neutrinos, gamma rays, and multiple electron-positron pairs. The muon fragments decay soon after, widening the spray of debris, as the shower makes its way to lower altitudes. The most important piece of debris is the neutron, which immediately slams into the most common constituent in the atmosphere, nitrogen-14, at extremely high speed. The free neutron is not captured by the nitrogen-14, in the classical way, becoming nitrogen-15. Instead, the collision causes a proton to eject from the nucleus, as the neutron sticks to it, and the result is a new atom of carbon-14.

This process occurs at high altitude, from 30,000 to 50,000 feet (9,144 to 15,240 km), and the carbon-14 readily mixes with the thin air at this height, combines with the free oxygen in the air, and becomes carbon dioxide, which is absorbed by plant life on the ground. The carbon-14 makes its way to most animal life through the consumption of plant life. The entire loading of carbon-14 in the atmosphere is stable at 70 tons (63.5 metric tons).

While we are largely shielded from the worst effects of cosmic rays by the magnetic field and by the thick overhead atmosphere, this radiation source becomes a serious consideration for space travel, where there is nothing to protect against the high-speed particles but the thin skin of a spaceship, built to be as light as possible. The background radiation on the ground is about 240 millirems (2.4 mSv) per year, with about 30 millirems (0.3 mSv) due to cosmic rays. In interplanetary space, the

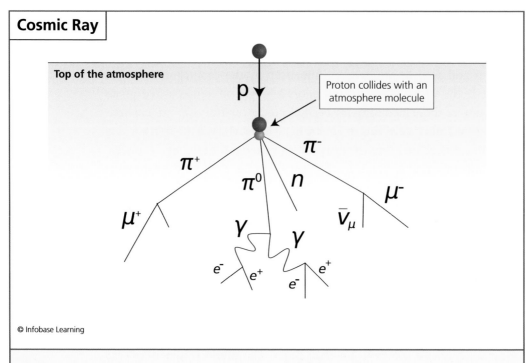

Cosmic Ray

An illustration of a cosmic ray from deep space hitting the upper atmosphere and causing a cosmic ray shower. Single particles split into many radiation events in the air, multiple times. A single proton (p) hitting an air molecule can cause positive, neutral, and negative pions (π⁺, π⁰, π⁻) and neutrons (n) to shower down. Further particle decays can emit positive and negative muons (μ⁺, μ⁻), gamma rays (γ), electrons (e⁻), positrons (e⁺), and muon neutrinos (ν_μ).

exposure would be 40 to 90 rems (400 to 900 mSv) per year from pure cosmic rays, and shielding against them is difficult. A 30-month mission to Mars could result in a 114 rem (1.14 Sv) exposure. The maximum career limit for radiation exposure for a worker in low Earth orbit (LEO), such as an astronaut on the International Space Station (ISS), is 100 to 400 rems (one to four Sv), as set by the National Council on Radiation Protection & Measurements (NCRP), and this ultimate limit is approached by a single trip to Mars.

Long-term exposure to ionizing radiation, or slight exposures every day for months at a time, can cause health problems, primarily the onset of cancer. This and other effects are discussed fully in the next chapter.

2 The Effects of Radiation

In chapter 1, quantities of radiation were mentioned, with radiation energies and intensities used to convey the relative serious natures of various types of emissions. The reader should not feel singled out if this is bewildering. Human beings are not aware of radiation levels, with the possible exception of sunlight, and there is no inherent feeling for safe and unsafe levels. When radiation was first discovered near the end of the 19th century, there was no human experience whatsoever in dealing with radiation intensities or dosages, and it would take 50 years to fully grasp the effects of this invisible force and to come up with safety limits. Before these limits could be established by observation, there first had to be a system of units to measure radiation dosage, the rate of dosage, and the ultimate radiation flux from a source.

This chapter begins with an explanation of the many ways of expressing a radiation measurement and what this means to the safe use of radiation. The chapter then moves on to describe the ways that different radiation types interact with matter, as was mentioned in chapter 1 while describing the radiation types. Human responses to radiation will be covered in more depth in chapter 7, *Medical Treatment of Radiation Poisoning.* A sidebar gives an example of continued monitoring of the effects of large impulse radiation doses and long-term, low-level exposure on the survivors of the *atomic bomb* drops at Hiroshima and Nagasaki, Japan, at the end of World War II.

UNITS OF RADIATION MEASUREMENT

The units of radiation measurement pertain to ionizing radiation, or electromagnetic radiation above ultraviolet rays on the spectrum, plus charged particles traveling at speed. Neutron radiation has no distinctive measurement notation, and its effects on human beings are usually measured in terms of secondary effects that cause ionizing radiation. Although radiation units based on the ionization effect were proposed as early as 1905, a formal system of measurement was not established until the meeting of the Second International Conference of Radiology in 1928 in Stockholm, Sweden. The first units to be used were the *curie* and the roentgen, named for early pioneers in nuclear physics.

The most fundamental unit of radioactivity is the curie (Ci), a measure of the total activity of a radiation source. It is equivalent to the activity of one gram of radium-226. A gram of radium-226 decays at a rate of 3.7×10^{10} disintegrations per second. One curie is a great deal of radiation, and for practical use with reasonable amounts of radiation it is usually cut down to the microcurie, 2.22×10^6 disintegrations per minute.

Care must be exercised when using the curie as a measurement of radiation. The curie is a unit of disintegrations per second, and not rays or particles per second. One disintegration of radium-226, for example, does not result in an ejected particle with a known damage potential. It results in the ejection of an alpha particle with multiple possible energies and at least one gamma ray, also with a range of possible energies and resulting variable penetration and damage potential. The curie is not directly measurable with a handheld instrument, such as a Geiger counter. A radiation probe cannot measure all radiation coming from a source. A source typically emits particles in random directions, and a probe measures only those particles that happen to be aimed directly at it and actually strike the instrument. There is also the problem of shielding. A probe that counts gamma rays will not necessarily count the less penetrating alpha particles, and every gamma ray that hits the probe will not necessarily be counted. There is only a probability that it will be counted. For these reasons, the number of curies of radiation from a source is not known from its measured activity but from its nuclide type and its mass. Each nuclide has known disintegration modes and decay rate, and the amount of radiation from it depends on how much of it is in the source. Any radiation-measuring device, such as the Geiger counter, therefore does not measure radioactivity, and it is not calibrated to read out on a scale of curies. A radiation instrument measures the amount of radiation hitting the probe and not the total number of particles per second coming from the source of radiation.

The official Système international d'unités (SI) unit of radioactivity is the *becquerel (Bq)*. It is named for Henri Becquerel, who shared the Nobel Prize in physics with Marie Curie for having discovered and characterized radiation. It is simply the number of disintegrations per second in a radioactive source, with no reference to a specific source. One curie is therefore 3.7×10^{10} becquerels, and one becquerel is 2.7×10^{-11} curies.

As is the case for any SI unit, becquerels may be prefixed to indicate exponent multiples of 1,000. One becquerel is an unrealistically small unit, and prefixes are commonly used with becquerels. One thousand becquerels is the kilobecquerel (kBq), or 10^3 becquerels. One million becquerels is the megabecquerel (MBq), or 10^6 becquerels. One billion becquerels is the gigabecquerel (GBq), or 10^9 becquerels; 1 trillion becquerels is the terabecquerel (TBq), or 10^{12} becquerels; and at the upper limit of practical radioactivity, 1,000 trillion becquerels is the petabecquerel (PBq), or 10^{15} becquerels.

The amount of the nuclide potassium-40 in the average human body disintegrates 4,000 times per second, with a radioactivity of four kilobecquerels. It is estimated that the detonation of a nuclear weapon in Hiroshima, Japan, in 1945 produced a radioactivity of 8×10^{24} becquerels.

Although the becquerel is the official unit of radioactivity, the curie is still used extensively. Calibration sources for radiation instruments, for example, are designated in microcurie units, and federal regulations controlling the possession of radioactive sources for industrial, research, medical, or instructional purposes are specified in terms of curies.

The roentgen (R) is a measurement of the effect of ionizing radiation. Specifically, one roentgen of ionizing radiation strips off one outer electron from each of 2.08×10^9 atoms, or the molecules in one cubic centimeter of dry air at standard temperature and pressure. The roentgen measures the effect of radiation on matter, removed from the source of the radiation, and it can therefore be measured with any number of radiation instruments. At a given distance, a greatly radioactive source causes more ionization than a slightly radioactive source, but the roentgen expresses a collective effect of the radiation and does not characterize the source. Roentgen is a measure of integral effect, or *dosage*, of radiation, whereas the curie is a measure of the rate of decay of a radiation source. It is the amount of radiation that has been absorbed over a period of time, where the effect of this radiation is ionization. It is possible for a radiation particle to pass through matter without causing ionization, as a small but real probability, and radiation that has no ionizing effect is not considered part of the dosage.

The roentgen, for practical measurements, is typically prefixed with milli (m), meaning 10^{-3}, or 1/1000th. An average radiation background dose for a human being is 300 millirems per year. A dosage of 500 rems in five hours of exposure to radiation is considered lethal for humans.

The roentgen was considered an SI unit until 2006, as long as it was differently defined as 2.58×10^{-4} coulombs/kilogram of air ionization at standard air density. It has officially been replaced by the *gray (Gy),* but it has been used for many decades, with thousands of papers written using it as a dosage unit, and it remains in use because of this familiarity.

An important variation of the roentgen as a unit of radiation dosage is the *roentgen equivalent man (rem).* The roentgen is a one-dimensional measurement, and while it is a good indication of the effect of radiation on matter in general, it does not address the subtleties of radiation effects on living tissue. The problem with the simple roentgen is that one roentgen of radiation has a different ionization effect, and therefore ultimate damage, on air than it does on a human being. Moreover, the effect of a given radiation on all tissues, from bone to blood, is different. Different types of ionizing radiation have different effects, as do different energies of the same radiation. Individual people can be affected in different ways by the same radiation dose.

The measurement of a radiation dose to an individual for the purpose of evaluating or predicting the effects of this dose is complex. To get a precise reading of radiation damage on each person would require a different unit of measure definition for everyone. Instead, there has been established an equivalent man, or a theoretical average human being, for which radiation dosage is standardized. In fact, for this hypothetical person, there is a standard liver, a standard spleen, a standard brain, and so forth, each defined as a specific sphere of polyethylene plastic from which standardized, calibrated radiation dosage measurements can be collected by putting a radiation probe in the center of each organ model. From such measurements, a weighting factor has been established, so as to adjust a roentgen measurement to reflect the potential damage to a standard human being. The factor is 1.07185. One rem equals 1.07185 roentgens.

Although the use of rems for radiation dosage is strongly discouraged, they have been used for so long it will be difficult to replace them completely with the newer SI unit, the *sievert (Sv).* They are commonly used to express everything from X-ray dosage to patients in hospitals to federally mandated human dose limits.

All dosage measurements are expressible as integral dose, or the total amount of radiation absorbed, and dose rate, or the speed with which a dose is absorbed. In industries involving radiation, a dosimeter measures an individual's accumulated dose, while a handheld radiation-measuring instrument commonly measures the dose rate. At a high rate, the dose adds up quickly, allowing less time in a radiation environment. A low rate means that the dose accumulates slowly, allowing more time at the task.

As a very rough rule, a dose of 10 rems, accumulated over the passing of one hour, is not harmful. A dose of 100 rems taken in one hour will cause sickness but rarely has a lasting effect. A dose of 1,000 rems or more in one hour will kill. At a background dose rate of 200 millirems, or 0.2 rems, per year, it would take 50 years to accumulate 10 rems from the various environmental radiation sources. The rate of radiation exposure is a factor in its effect on humans. A dose of 100 rems in one hour may have serious health effects, but 100 rems distributed over the course of 10 years will probably be undetectable. Radiation sickness does not start suddenly after the accumulated dose reaches 100 rems over 10 years.

An alternate unit by which radiation dose may be measured and evaluated is the *rad*. While the roentgen is a measure of the amount of ionizing caused by radiation, the rad is a unit of absorbed energy, independent of any ionization. The rad was developed for use in expressing quantities of X-rays used in radiation treatment of cancer tumors and was first suggested in 1918. One rad was defined as the quantity of X-rays that would destroy a specific number of cells in a model malignant tumor. The unit was more precisely defined in 1953 as the dose of radiation causing the absorption of 100 ergs of energy in one gram of matter. In 1970, the definition was restated for the benefit of the SI as being the dose causing 0.01 *joules* of energy to be absorbed per kilogram of matter.

The SI unit that has replaced the roentgen equivalent man, or rem, as an adjusted measurement of biological dose is the sievert, named for Rolf Maximilian Sievert (1896–1966), a Swedish medical physicist who contributed much to the study of the biological effects of radiation. The rad as a measure of the purely physical aspect of radiation exposure, or absorbed energy, was replaced in 1975 by the gray, named for Louis Harold Gray (1905–65), the British inventor of radiobiology. The plural of gray is also gray.

The definition of the sievert is one absorbed joule of radiation energy per kilogram of matter, adjusted for biological effects by two factors: Q,

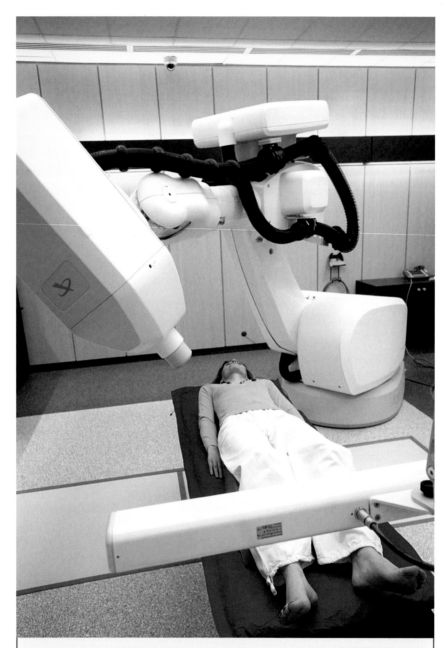

The treatment of cancer by robotized radiotherapy at the Centre Oscar Lambret, regional center in the fight against cancer in Lille, France. A radiation source is being aimed precisely at a tumor in a woman's neck to destroy cancerous tissue without affecting healthy tissues. *(Centre Oscar Lambret/Photo Researchers, Inc.)*

depending on the radiation type, and *N,* depending on the organ irradiated, the species of the irradiated individual, and the time and volume over which the dosage is received. These two factors are combined into the radiation weighting factor, or W_R. In 2002, it was decided that the N factor creates too much confusion, and the SI has since deleted it from the definition of the sievert. All mentions of sievert measurements prior to 2002 consider the N factor.

The weighting factor is now dependent only on Q, or the relative biological effectiveness, of a radiation type. For all electromagnetic radiation, beta rays, and muons, Q = 1. For alpha particles, Q = 20, and this is a clear demonstration of the damaging effect or the danger from alpha particles. Alpha radiation is 20 times worse than X-rays or gamma rays, both of which are electromagnetic in nature. The Q factor for neutrons is entirely dependent on the energy or speed of the particle. Neutrons at 100 KeV (kilo electron volts) are the worst at Q = 20. Neutrons of energy less than 10 KeV have a Q = 5, as do neutrons of an energy greater than 20 MeV. Neutrons between 10 and 100 KeV, or between 2 MeV and 20 MeV, also have a Q = 10. One joule of energy from 100 KeV neutrons absorbed by one kilogram of material results in 20 times more radiation damage than is caused by the same energy deposited by gamma rays.

The N factors, while officially obsolete, are still interesting for what they express about the relative sensitivity of various organisms and the organs in a human body. For a standard human being, N = 1. For viruses, bacteria, and protazoans, the factor is at most 0.03 and can be as low as 0.0003, indicating that it is much less likely to cause harm to a single-cell organism with the same radiation dose. Some plants, on the other hand, can have an N = 2, meaning they are twice as sensitive to radiation as human beings, but they can also be as low as N = 0.02. Pine trees seem exceptionally sensitive to radiation. Reptiles are almost equal to humans, with an N = 1, but some can be as low as N = 0.075. Mollusks can be as low as N = 0.006 but never higher than N = 0.06. Of all multicell creatures, insects seem to have the most resistance to radiation damage, with a minimum N = 0.002 and a maximum N = 0.1. The N factor for birds is between 0.6 and 0.15, depending on the species.

For a human being, the N factor broken down by organs exposed to radiation shows the relative sensitivities of the different tissues. The skin and dense bone has an N = 0.01 factor, which is fairly low, but the bone marrow, as well as the colon, lung, and stomach, have an N = 0.12 factor. The gonads are the most sensitive, with an N = 0.20 factor. The bladder, brain, breast, kidney, liver, muscles, and the remaining tissues are insensitive in comparison, with an N = 0.05.

For practical radiation exposure measurements, the millisievert (mSv), or 0.001 sievert, and the microsievert (μSv), or 0.000001 sievert, are commonly used. As an approximation of the feelings caused by various doses of radiation expressed in sieverts, one sievert causes nausea. An accumulated dose of two to five sieverts causes hair loss and hemorrhage and will cause death in many cases. More than three sieverts will cause death in 50 percent of radiation absorption cases within 30 days. If more than six sieverts of radiation are absorbed, survival is highly unlikely.

The collective dose to a population is expressed in man-sieverts (manSv), which is the number of people exposed times the average exposure in sieverts. (Note that the sverdrup, a unit of volume transport used in oceanography to measure the transport of ocean currents and named for Harald Sverdrup, is also abbreviated Sv.) One sievert is equal to 100 rems.

One gray, the SI replacement for the rad, is defined as one joule of radiation energy absorbed in one kilogram of matter, and one gray is equivalent to 100 rads. If the ionization energy of air is about 36.2 joules per coulomb, then the energy transfer of one gray is equivalent to the ionization of about 115 roentgen.

There are 20 formal SI prefixes to express multiples of 10 gray, ranging from the yoctogray, or 10^{-24} Gy, to the yottagray, or 10^{24} Gy. For practical purposes, there are: the kilogray (kGy), or 1,000 Gy; the milligray (mGy), or 0.001 Gy; and the microgray (μGy), or 0.000001 Gy.

The average radiation dose received from an abdominal computed tomography (CT) scan is 8.0 mGy and from a pelvic CT scan 25 mGy. A selective spiral CT scan of the abdomen and the pelvis is 30 mGy. An abdominal X-ray is only 1.4 mGy. A whole-body exposure to about five gray, or 5,000 mGy, at one time usually results in death within 14 days. Absorbing five gray of radiation over a lifetime results in no measurable effect. To express exposure in gray takes into account no biological factors, unlike the sievert or the rem.

Any radiation dose measurement is also measurable as a rate of exposure, such as rem per hour or sieverts per hour, and the rate of dosage is important when scheduling radioactive work or diagnostic tests. If the allowed daily dose of a radiation worker is five millirems and the measured dose rate inside a room from a radioactive source is one millirem per hour, then the worker may not be in that room for more than five hours in a day.

IONIZATION AND ENERGY DEPOSITION

The use of the rad or the gray in specifying radiation dose is applicable to both ionizing radiation and neutrons, both of which are moving, energetic particles that are stopped by colliding with matter. In being stopped or at least slowed down by interaction with solid objects, radiation will lose energy in the transaction and will heat up the target. The dose received depends entirely on the nature of the obstacle and the radiation. Gamma rays, for example, will lose all energy in a lead shield, and the lead will therefore absorb a maximum dose in rads. Neutrons, on the other hand, can pass right through lead and continue at speed, leaving little energy in lead and therefore subjecting it to a minimum dose. A light substance, such as paraffin, is completely different, not stopping gamma rays but taking neutrons down to thermal speed and therefore absorbing a maximum dose. Living soft tissues are much closer to paraffin than to lead in radiation interactions, just as dense, hard bone is closer to lead. An example of this difference can be seen in a medical X-ray film, where the stark contrast between bone and soft tissue is created because of the relative difference in radiation stoppage. X-rays projected onto a person's arm can make it through soft tissues and expose the large sheet of photographic film underneath the limb, while the two long bones capture some X-rays, denying them exposure of the film. The result is a photographic shadowgraph, showing structure in the arm that visible light, an electromagnetic ray that is lower down on the spectrum, is unable to penetrate and reveal.

Electromagnetic radiation of a sufficient energy, starting with ultraviolet rays, is considered to be *ionizing radiation* because of its ability to strip off an electron from an atom and leave it with a net positive charge. An atom with a missing electron is a positive ion. A most noticeable effect of ionization is sunburn, and it can also occur from overexposure to X-rays and gamma rays, which are above ultraviolet rays on the energy and frequency spectrum.

The high-energy particle mode of gamma rays, named photons, can interact with matter in one of three ways: the photoelectric effect, Compton scattering, and pair production.

The photoelectric effect, discovered by the German physicist Heinrich Hertz in 1887, occurs at lower energies, starting at a few electron volts, and can be the result of visible light. If visible light impinges on the surface of an alkali metal, such as cesium, then the light is absorbed and electrons are emitted, ionizing the surface of the metal target. In this case of low energy electromagnetic radiation, light is considered not as a hard, massive particle bouncing electrons out of the way but as a wave of energy that is fully absorbed by the target. Electrons leave the target because of excess, imparted energy, and not because of a billiard-ball collision. Nonmetallic targets require a greater energy than visible light is able to exert for electrons to leave the surface, beginning with the upper end of the ultraviolet sub-spectrum. The photoelectric effect is used in television cameras and digital cameras to capture scenes in visible light.

The American physicist Arthur Compton (1892–1962) discovered in 1923 that a high-energy gamma or X-ray photon will lose energy when it

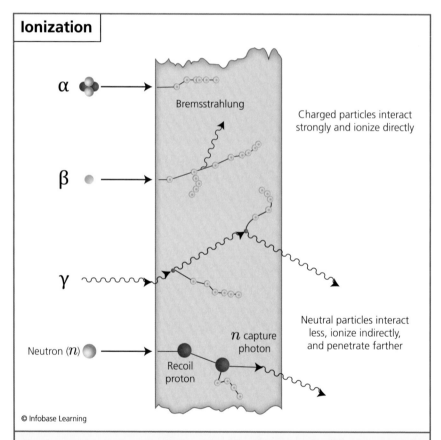

Ionization

Bremsstrahlung

Charged particles interact strongly and ionize directly

Neutral particles interact less, ionize indirectly, and penetrate farther

n capture photon

Neutron (n)

Recoil proton

© Infobase Learning

The four major types of radiation interact with matter in different ways. An alpha particle (α) ionizes atoms by heavy collision, bouncing the target atom forward. A beta ray (β), or traveling electron, emits a secondary X-ray, or bremsstrahlung, as it is stopped by matter. A gamma (γ) ray reflects off atoms by being absorbed and then reemitted, losing energy with each reflection. A neutron (n) can collide with a hydrogen atom and exchange momentum with it or it can be absorbed by a nucleus and cause it to become radioactive.

collides with an electron. The electron hit by the photon will increase in energy, and this indicates that a photon acts as a traveling particle, having mass and momentum, and it can influence a standing electron. The energy of an X-ray or gamma ray that produces Compton scattering is on the order of 1,000 electron volts, or a KeV. Compton scattering of an electron in the outer orbit of an atom results in ionization.

Electrically charged particles, such as the alpha particle and the two types of beta particle, strongly ionize matter through which they are traveling. The alpha particle is particularly damaging, because it is doubly charged, with two positively charged protons in its structure. Its strong interaction with passing atoms in a material also means that it has a very short range. Alpha particles can be stopped by a sheet of tissue paper or simply by a centimeter of air. A stopped alpha particle is simply an ionized helium atom. Enclose a strong alpha-emitting radiation source in an evacuated glass tube, and the tube will fill with helium gas. Helium gas trapped underground, typically in Texas oil wells, began as alpha particles emitted from radioactive nuclides miles underground, bubbling slowly up through microscopic cracks in the rock and magma in and below Earth's crust.

An energetic alpha particle can knock an orbiting electron out of an atom with such force that it becomes a secondary radiation, named a *delta ray* by the English physicist J. J. Thomson (1856–1940) in 1905. The delta ray is also called the knock-on electron. It acts as a β– particle, or beta negative.

The beta particle, with one positive or negative charge, interacts less strongly with matter than the double-charged alpha and therefore travels farther before it is completely stopped. A beta particle can be stopped, or shielded against, by a thin sheet of metal. A pie tin protects against beta rays. A beta ray traveling through air or vacuum that hits a metal surface will produce naturally occurring X-rays, or *bremsstrahlung,* a secondary effect of beta radiation.

Although the neutron is a neutrally charged particle that can cause no direct ionization, there is a secondary phenomenon. When a high-speed neutron crashes directly into a hydrogen atom, the two particles do a perfect momentum transfer, like a cue ball fired into a sitting eight ball. The neutron comes to a stop, and the hydrogen nucleus, which has almost the same mass as the traveling neutron, shoots off at the previous speed of the incoming neutron. The result is a positively charged energetic particle, or proton. It cuts a stream of ionization through matter similar to that of a beta particle, finally stopping when all its initial energy has been expended in the ionizing processes. It has about the same range in matter as the beta minus particle.

MOLECULAR MODIFICATION

Radiation can cause chemical reactions, either breaking existing chemical bonds or causing atoms or molecules to bind into more complex molecular structures. In the process of *radiolysis,* for example, water is broken down into hydrogen and oxygen by the action of ionizing radiation. In nuclear reactors using water as a coolant, the circulating water must be run through a catalyst pack, to recombine those water molecules that were broken as the water was pumped through the high radioactivity of the reactor core.

Organic polymers are also sensitive to radiation. Ionizing radiation can convert monomers to polymers, cross-link adjacent polymers, and break polymer chains. There are industrial uses for these phenomena. Heat shrink tubing, used for everything from product packaging to electrical cables, is made by cross-linking a thermoplastic material using beta rays.

Molecular effects in living tissues are more complex. The harmful results of acute radiation exposure appear to be due to the ionization effect, which destroys or damages some of the constituents in a living cell. The products formed by the molecular changes may act as cell poisons, as a secondary effect. Among the observed consequences are the breaking of chromosomes, swelling of the nucleus, swelling of the entire cell, increasing the viscosity of cell fluid, and increasing the permeability of the cell wall. The process of mitosis, or cell division, is delayed in time due to radiation exposure. In cases of larger radiation doses, mitosis cannot occur, and the replacement of damaged cells becomes impossible.

ACTIVATION

Neutrons, and in some cases alpha particles, can induce radioactivity in nonradioactive nuclides and cause radioactive decay into another element. For example, if a common nuclide of silver, silver-109, absorbs one neutron, it becomes silver-110. Silver-110 is neutron-heavy, meaning it is unstable, and it has a half-life of 24.6 seconds. It undergoes a $\beta-$ decay, meaning that the extra neutron becomes a proton, and an electron comes shooting out. With the neutron converted into a proton, the silver-110 becomes cadmium-110, which is perfectly stable.

Neutron activation is based entirely on probability. The probability of an activation depends on the nuclide involved in the collision and the energy, or speed, of the incoming neutron. This probability of activation, or of any neutron-nuclide interaction, is expressed as the effective cross section of the nuclide, dependent on the neutron energy. A large cross section means that the nucleus is easy to hit. A small cross section reduces the probability of hitting it with a neutron.

ACUTE RADIATION SICKNESS

The clinical name for radiation sickness is acute radiation syndrome (ARS), as defined by the Centers for Disease Control. It is a form of tissue and organ damage caused by a large dosage of radiation accumulated over a short time period. This condition is rare, as it is caused by proximity to nuclear warfare, uncontrolled assemblies of fissile materials, or fissile material processing plant accidents. The symptoms for an average or standard human being vary with the dose level, as follows:

5–20 rems (0.05–0.2 Sv)
No detectable symptoms. Any actual symptoms are hidden in the noise of normal human health issues.

20–50 rems (0.2–0.5 Sv)
The white blood cell count decreases temporarily.

50–100 rems (0.5–1 Sv)
A mild feeling of sickness with headaches. The risk of infection is increased due to a disruption of the immune system. Temporary male sterility is possible.

100–200 rems (1–2 Sv)
Mild to moderate nausea with occasional vomiting, followed by a latent phase of 10 to 14 days, when the nausea symptom seems to have disappeared. After the latent period, a general feeling of illness and fatigue appear. The immune system is depressed, and there is a risk of infections. Temporary male sterility is probable. In pregnant women, spontaneous abortion or

stillbirth will occur. All symptoms are probabilistic, with a 50 percent probability of these effects with a 200 rem dose. This dose is considered light radiation poisoning, and it is fatal after 30 days of sickness in 10 percent of cases.

200–300 rems (2–3 Sv)

Nausea onsets beginning three to six hours after acute exposure and lasting for up to a full day. Following the initial nausea is a seven- to 14-day latent period. After the latency, during which there are no symptoms, there is hair loss all over the body, general illness, and fatigue. The immune system is virtually destroyed, with a massive loss of white blood cells. Infection is a great risk. Permanent female sterility is possible. Convalescence is necessary for at least one month. This is considered moderate radiation poisoning, and it is fatal in 35 percent of cases after 30 days of sickness. At the upper end of the scale, 300 rem, nausea is 100 percent guaranteed and general illness is a 50 percent probability.

300–400 rems (3–4 Sv)

In addition to the symptoms listed above of 200–300 rem exposure, there is uncontrollable bleeding in the mouth, under the skin, and in the kidneys after the latent period. Illness caused by this dosage is considered to be severe radiation poisoning. There is a 50 percent chance of fatality after 30 days of illness. The characteristic symptoms are 50 percent probable with a high-end 400 rem (4 Sv) dose of radiation.

400–600 rems (4–6 Sv)

All the symptoms of a 300–400 rem exposure start an hour to two hours after exposure to radiation and last for up to two days. After the initial symptoms, there is a seven- to 14-day latent period in which there are no symptoms. After the latent period, the same symptoms reappear with enhanced intensity. Convalescence takes up to a year. Female sterility is likely. This is acute radiation poisoning, and it is fatal in 60 percent of the cases after 30 days. The cause of death is usually internal bleeding and uncontrollable infections after two to 12 weeks.

600–1,000 rems (6–10 Sv)

Gastric and intestinal tissues are severely damaged. Symptoms begin 15 to 30 minutes after radiation exposure and last for up to two days. The latent period is shortened to five to 10 days, after which death occurs with nearly 100 percent certainty. Possible survival depends on intense medical care, including a bone marrow transplant. The bone marrow is completely destroyed, and intestinal and gastric tissues are severely damaged. Possible recovery would take several years and would probably never be complete.

1,000–5,000 rems (10–50 Sv)

This exposure is 100 percent fatal after seven days. Spontaneous symptoms occur within five to 30 minutes. Chemical receptors in the brain are directly activated, and this causes a feeling of powerful fatigue and immediate nausea. Several days of comparative well-being follow. This latency period is called the walking ghost phase. After the latency, severe dehydration leads to chemical imbalance and death. Cell death in the gastric and intestinal tract causes massive diarrhea and intestinal bleeding. Death occurs with delirium and coma due to a breakdown in circulation. There is no treatment for this condition. Conditions for this degree of exposure have depended on touching proximity to a supercritical sphere of naked plutonium.

More than 5,000 rems (50 Sv)

There are two documented cases of exposure to more than 5,000 rems. A worker at the Wood River Junction fuel-processing plant in Rhode Island received an exposure of 10,000 rems (100 Sv) on July 24, 1964. He survived for 49 hours. On September 30, 1958, a worker at a fuel-processing facility at Los Alamos, New Mexico, was exposed to 18,000 rems (180 Sv). He died after 36 hours. Both incidents were criticality accidents in which impromptu nuclear reactors were accidentally assembled with a person in close proximity. Firefighters at the Chernobyl reactor disaster in 1986 may have received doses greater than 30,000 rems (300 Sv), but there was no instrumentation in use that could measure this extreme dose. All died.

LESSONS LEARNED AT HIROSHIMA AND NAGASAKI

The two nuclear weapons dropped on Japan in August 1945 had immediate and long-term profound effects. They were the stimulus behind the unconditional surrender of Japan after seven years of aggression. They prevented the Allied invasion, saving hundreds of thousands of lives in the Allied armed forces. However, the two bombs also represented a scientific experiment of horrible consequences. For the first time, a huge population was exposed to radiation in a wide spectrum of dosage, dose duration, and dose rates. In all, taking into account delayed effects, about 220,000 people died.

In 1945, surprisingly little was known about the effects of radiation on living things. The only experience with radiation injuries was from an occasional overuse of medical X-rays and from the reckless use of radium-226 as an industrial material and as a medical therpeutic component. The X-ray injuries gave scientists a rough idea of the damage caused by low-energy ionizing radiation, and horrific deaths from exposure to radium in the luminous watch dial industry gave a hint as to the power of alpha particles. There was, however, no experience with the mixed set of radiations from atomic bombs, ranging from infrared light to high-speed neutrons, with exposure times ranging from the millisecond pulse of the detonations to decades of heightened background radiation. The wartime populations of Hiroshima and Nagasaki provided a large sampling, and studies of the effects of the atomic bomb radiation continue to this day.

As the nuclear weapons were being developed at Los Alamos, New Mexico, during World War II, not much effort was expended on predicting the radiation effects on enemy personnel. It was assumed that most casualties would be from the mechanical effects of the extreme blast of the bomb, exploded in the air at an altitude of 1,000 feet (305 m), and this was not an entirely wrong estimate. Approximately 30 percent of the casualties from the atomic bombs were radiation related, but the

RADIATION-INDUCED CANCER

Radiation exposure can induce various malignancies, usually following a latency period of 20 to 40 years. Melanoma, for example, can be induced by exposure to the ultraviolet ionizing radiation in sunlight. The most probable radiation cancers are basal cell carcinoma, followed by squamous cell carcinoma. These diseases are in the general category of skin cancer. The risk of cancer by radiation exposure is approximately lingering effects of radioactive fallout and delayed biological effects from the initial radiation pulse were not anticipated.

The most immediate effect of the bomb radiation was the flash burn, caused by ionizing radiation from the nuclear fireball, which was more than 600 feet (183 m) in diameter. Flash burns were fatal to nearly all people in the open, with nothing to shield them from the radiation, within 6,000 feet (1,829 m) of the detonation center of the bomb projected onto the surface of the city, or ground zero. As far out as 14,000 feet (4,267 m), burns were sufficiently serious to require medical treatment.

The radiation burst lasted about one second. The severity of burns ranged from erythema, or reddening, very similar to a common sunburn, to charring of the outermost layers of the skin. At the nonlethal distance from the explosions, the initial reddening was followed by the development of a walnut coloration of the skin, referred to as the mask of Hiroshima. Infection of severe burns was likely, due to suppressed immune systems, and thick scars, or keloids, were likely to form over burned areas.

At distances up to 10 miles (six km), those individuals who happened to be looking at the sky where the bomb exploded experienced permanent retinal burns. The degree of flash burn seemed to depend on what was between a person and the bomb. Behind any solid object, a burn was much less likely, even at a relatively short distance. Even the type of clothing worn could make a difference. Women wearing patterned cloth were subjected to patterned burning, with light-colored cloth reflecting away ionizing photons but dark patterns absorbing and conducting the energy to the skin.

In 1948, President Harry S. Truman directed the establishment of the Atomic Bomb Casualty Commission to conduct investigations into the lingering effects of radiation among the survivors of the Hiroshima and Nagasaki bomb attacks. As of March 23, 2008, there were 243,692 people recognized as living survivors of these incidents, and approximately 1 percent have radiation-induced illnesses. In a careful accounting of the survivors, 231 have died from leukemia and 334 from solid cancers attributed to bomb radiation, including radioactive dust fallout following the detonations.

proportional to the integral dose received, although the incidence of disease due to very low dose is difficult to quantify. Unusually large dosages, such as those incurred in atomic warfare, have been shown to produce leukemia and general cancers in solid flesh.

Radiation was first linked to cancer in 1902, as X-ray technicians and radiologists were observed developing cancers on hands and body surfaces routinely illumined in X-ray beams. With the decoding of the DNA molecular structure in 1953 by James D. Watson (1928–   ) and Francis Crick (1916–2004), the effect of ionizing radiation on this complex, fragile polymer became clear, and the mechanism of radiation-induced cancer was the first cause of malignancy that was clearly understood. When radiation wrecks a DNA molecule, the damage is usually repaired, but if irreparable, it leads to cell death, or apoptosis. It is possible for a cell to experience a nonlethal, botched repair, and this becomes an inappropriate mutation of the genetic code. The surviving DNA molecule is able to duplicate itself in the normal way, producing abnormal cells, and this process is defined as cancer.

A controversial observation concerning low-level radiation dosage is that chronic low doses of ionizing radiation reduce the impact of a subsequent large dose of radiation. This hypothetical mechanism is *hormesis.* The theory is that low levels of radiation, near the level of common background radiation, help immunize cells against nonlethal DNA damage by keeping the DNA repair procedure active, causing higher levels of cellular DNA repair proteins to be available at all times. The statistical noise caused by all other causes of cancer, such as cigarette smoke, prevent experimental confirmation of this assertion. Regulatory bodies in charge of monitoring and setting limits on occupational radiation exposure base dose limits on a simple linear no-threshold (LNT) model of dosage, stating that the risk of cancer induced by radiation is directly proportional to the dose of radiation received, with no defined lower cutoff dose. If there is error in this assessment, it is on the side of safety.

The concept of radiation in the environment, put there by both nature and by human enterprise, is further expanded upon and discussed in the next chapter. Radiation and the effects of radiation, large and small, are all around us, but the surface of Earth may be the safest place in the universe to stand.

3 Radiation and the Environment

For millions of years, as humankind has evolved and developed from hunter-gatherers on the plains of Africa to the spread of enterprise all over the earth, humans, animals, and plants were being bombarded from all directions with radiation of most types, from ionizing electromagnetic to high-speed particles. The exception in this range of radiation types may be neutrons, to which very few people have ever been exposed. This radiation dosage did not seem to hinder the advancement of plant and animal life. It may, in fact, have encouraged it.

This chapter lists the various sources of naturally occurring and artificial radiation in the environment, giving the impression that an individual's dose rate at a given time depends largely on where that person is standing. Some places on Earth are better locations than others, from a radiation-absorption standpoint. This background radiation is generally so slight, of such a small dose rate compared with X-ray machines or active nuclear reactors, it causes no immediate health risk. Still, as consciousness of radiation and its effects became clearer in the 20th century, scientific interest in background radiation increased. It is now known that sunlight, the largest source of environmental radiation, is a major cause of skin cancer, so limits to exposure are now advised. Radiological ground surveys and cosmic ray observatories have focused attention on natural radiation sources and enhanced our perception of background radiation as a factor in human survival. The sidebar on pages 50–51 discusses the concept of radiation-induced gene mutation and its possible link to evolution.

WIDELY USED BUILDING MATERIALS AS A RADIATION SOURCE

One of the oldest and most widely used building materials is rock. Ever since the first humans decided to build something permanent, rocks have been used in all forms and sizes to build everything from temples to dams across rivers. Enormous stones, weighing several tons each, have been used, as have pea-sized gravel and even sand. Whatever size, most rocks in Earth's crust contain at least a detectable trace of a radioactive element.

The two radioactive elements most likely found in rocks are uranium and thorium. The most likely isotopes, or nuclides, of these elements to be found naturally are uranium-238, which is 99.28 percent of the uranium, and thorium-232, which is all of the thorium. Uranium-235 is barely present, at 0.72 percent of the occurring uranium, and there is a trace of uranium-234, at 0.0055 percent. These nuclides all undergo alpha decay, with subsequent gamma rays, and the resulting lesser nuclides, or daughter products, are all neutron heavy and radioactive. The half-lives of the natural nuclides are quite long, and this leads to very low decay rates. The half-life of uranium-238 is 4.47 billion years, uranium-235 is 704 million years, and uranium-234 is 244,000 years. The differences in these half-lives account for the relative abundance of the three nuclides. Uranium-234 has the shortest half-life, and most of it has decayed in the 4.5-billion-year life of Earth, while about half of the uranium-238 that was originally present when Earth formed is still here. Thorium-232 has a half-life of 14 billion years and is possibly the most abundant radioactive nuclide in Earth's crust.

Of these natural nuclides only uranium-235 is *fissile,* meaning that it can fission with the slight stimulus of a neutron capture, and that furthermore it can sustain a continuous state of *chain reaction* fission by releasing more neutrons than it absorbs. This state of chain reaction is physically impossible in geological formations, as the atoms of uranium-235 are too sparsely disbursed in rocks to form a critical mass. This has not always been the case. Uranium-235 decays faster than uranium-238, so its concentration in uranium-bearing minerals has gradually decreased to 0.72 percent. However, as recently as 1.5 billion years ago, the concentration was close to 3 percent, which is the concentration of uranium-235 in commercial nuclear power reactor fuel. Under certain conditions of groundwater intrusion, heavy subterranean concentrations of uranium ore were capable of nuclear criticality, and fission occurred. This unusual condition produced, as a by-product, plutonium-239, from uranium-238 capturing excess neutrons produced by fission. Aside from production in ancient geologic fission, the fissile nuclide plutonium-239 does not occur in detectable concentrations in uranium concentrations. At the Oklo uranium mine in the Central African state of Gabon, a concentration of more than 3,000 pounds (1,400 kg) of plutonium-239 has been identified from natural, sustained fission.

In roundabout ways, all these natural radionuclides eventually decay into stable, nonradioactive lead isotopes, but during the process much radioactivity is released. An example is the decay of thorium-232. Thorium-232 decays directly into radium-228, a highly radioactive nuclide with a 5.7-year half-life. This is an alpha decay, which means that the thorium-232 nucleus has lost two protons and two neutrons, thus reducing the nucleus weight by four, going from 232 to 228. The loss of two protons takes the atomic signature, or the element, down two steps on the periodic table of elements, completely changing the atom from thorium to radium.

The radium then decays into actinium-228. This is an entirely different decay mode. It is a β– decay, meaning that a neutron in the radium nucleus has transformed into a proton. This action ejects a fast-moving electron, and it kicks the atomic signature up one on the periodic table, making radium into actinium. The weight of the nucleus remains the same. It is still 228.

Actinium-228 has a half-life of only 6.1 hours, and it does another β– decay into thorium-228, retaining the nuclear weight. Thorium has therefore decayed, eventually, into thorium, but with a lower nuclear weight. This version of thorium has a 1.9-year half-life, and it alpha decays into radium once more. This is a different radium nuclide. It is radium-224, with a 3.6-year half-life. It undergoes alpha decay and becomes radon-220, a highly radioactive noble gas, with the short half-life of 55 seconds.

Radon-220 quickly decays into polonium-216 by alpha decay, and this nuclide of polonium has an even shorter half-life of 0.14 seconds. In a flash, it decays by alpha emission into lead-212. While the thorium has decayed into lead, as promised, it is a radioactive nuclide of lead and will not remain lead for long. It has a 10.6-hour half-life, and it executes a β– decay, upward into bismuth-212.

Bismuth-212 has a 61-minute half-life, but it can decay either by β– or by alpha, with the decay mode randomly selected. In the alpha decay mode, the bismuth-212 changes directly into thalium-208. In the β– mode, it changes into polonium-212. Both possibilities of a decay outcome become stable lead-208. The polonium-212 does so quite rapidly, with a half-life of 3×10^{-7} seconds, or 0.3 millionths of a second, for an alpha emission. The thalium-208 takes a bit longer, decaying by β– with a half-life of 3.1 minutes. The radioactivity given off by naturally occurring thorium-232, on its way to becoming stable lead, adds to six alpha particles and four high-speed electrons.

Granite, a widely occurring type of intrusive, felsic, igneous rock, is a commonly used building material, either as a cut stone block or as gravel in concrete. There are many variations of granite, and it can appear in any color from light pink to black. It also seems to have an unusually high radioactivity, due to trace inclusions of uranium or thorium. Granites generally contain about 10 to 20 parts per million of uranium, not in a recognizable rock component, but as a distributed contamination of uranium oxide. Granite is composed of various concentrations of potassium feldspar, plagioclase feldspar, quartz, muscovite, biotite, and hornblende-type amphiboles, none of which contain uranium. Limestones and other sedimentary rocks used as building materials, as well as tonalite, gabbro, or diorite, contain a lower uranium contamination, on the order of one to five parts per million. Other building materials, even wood and straw, all contain trace amounts of uranium and thorium, but usually in unimportant concentrations. A million pounds of granite could contain 20 pounds of uranium, making this material a small contributor to natural background radiation.

Villages located over granite bedrock, for example, could accumulate a higher dose of radiation than other communities. Wells drilled into granite are of some concern. The potential for harmful radiation dosage is not from the mineral itself. Uranium or any other radioactive metal that is firmly encased in granite or other rock is effectively shielded from emitting radiation. Only those few radioactive atoms that happen to be on or near the surface of an exposed rock are capable of hitting a biological unit with radiation, and even in this case the individual must be pressed against the rock to receive radiation. The concern over uranium and thorium decay is not directed at the original nuclide but toward a specific nuclide that comes about in the decay chain—radon. All nuclide forms of radon are radioactive, and it happens to be a noble, or inert, gas. Radon cannot combine chemically with anything, and it is thus free to wander as a gas under normal conditions of temperature and pressure. It can easily flow through microscopic cracks in rock, and it even migrates through the solid matrix of rock, although slowly. There is no containing it, and by random, thermal motion it will find its way out of the rock and into the atmosphere, where it can be breathed.

A deep well drilled in granite will accumulate and concentrate radon gas, as air currents will not carry it away. Similarly, an unventilated basement with a granite, bedrock floor will accumulate radon. A basement with a concrete floor and possibly within concrete or granite-walled house foundation will have a similar accumulation, because the gravel in the concrete is likely uranium-bearing granite. If radon is in the air, then radon can be drawn into the lungs. It will also be blown out of the lungs, but if in its brief stay it happens to decay, then it hits the lung tissue with a five or six MeV ionizing alpha particle.

The main hazard of radiation in building material is therefore not in being bombarded by radiation from it under foot, or finding it in food, or being burned by touching it, but by breathing one component of the complex decay chain from traces of radionuclides distributed in it. What saves humanity from death by radon poisoning is the short half-lives of the possible radon decay products, meaning that it is more likely to transform into a nongaseous, chemically reactive element before it has a chance to be sucked into the lungs. Radon-220 from thorium decay has a half-life of 55.6 seconds. From uranium decay, radon-218 has a half-life of 35 milliseconds, and radon-222 has a half-life of 3.8235 days.

RADIOACTIVE POTASSIUM AND CARBON

A half-life of 1.28 billion years means that about one-sixteenth of all the potassium-40 that ever existed on the planet Earth is still here. For every 100,000 atoms of potassium, there are 12 atoms of radioactive potassium-40. The average adult, weighing 154 pounds (70 kg), contains five ounces (140 gm) of potassium, of which 0.000596 ounces (0.0169 grams) is potassium-40. This amount of potassium-40 decays at a rate of 266,000 disintegrations per second, with 89 percent of the decays being 1.33 MeV $\beta-$ events and 11 percent being 1.46 MeV gamma rays. Virtually all of the beta decays are absorbed by the body and are not externally observable. About 50 percent of the gammas escape into the outside, and 50 percent are completely absorbed. The annual dosage from this internal radiation source is 18 millirads (0.18 mGy), or 16 millirads (0.16 mGy) from the beta rays and 2 millirads (0.02 mGy) from the gamma rays. An average human body is thus a source of 1.46 MeV gamma rays, exiting at 14,600 rays per minute in all directions.

In nature, potassium occurs only as an ionic salt and is found dissolved in seawater and as part of many minerals, including biotite, muscovite, metamorphic hornblende, and volcanic feldspar. It is necessary for the function of all living cells and is present in all plant and animal tissues. It is most important for the function of the brain and nerves, and it influences the osmotic balance between cells and the interstitial fluid that separates them.

Carbon-14 is also a radioactive nuclide present in all living things, but it is a smaller component. Carbon-14, produced by cosmic ray

ENVIRONMENTAL RADIATION AND GENE MUTATION

Externally induced mutations of the genetic material in living things is a known and accepted process. Ionizing radiation, as low on the energy spectrum as ultraviolet light, can change the chemical properties of base pairs in DNA, or the information-carrying rungs of the ladder structure of this highly complex molecule. Two of the nucleotide bases in DNA, cytosine and thymine, seem most vulnerable to radiation damage as the molecule is in the process of copying itself. Ultraviolet light, for example, can cause thymine bases in a DNA strand to pair with each other, creating a bulky "dimer," or a weakly connected string of two similar subunits.

Radiation-induced mutations can have different, randomly selected effects, depending on where exactly in the genetic material the damage occurred. The damage must be slight enough to allow the DNA molecule to manage an incorrect repair and successfully complete the replication process. The mutation can be neutral, meaning that it has neither a harmful nor a beneficial effect on the organism. It can be deleterious and decrease the fitness of the organism, or it can be advantageous and have a positive effect on the subject of mutation. There are also nearly neutral mutations, which instigate a change that is either slightly advantageous or slightly deleterious. Most nearly neutral mutations seem to be slightly deleterious.

Concern over the effects of radiation on human populations in the 1950s led to serious study of mutation effects. A working theory was that harmful mutations are eliminated from a mutating population more quickly by natural selection, while bombardment of the upper atmosphere, is only one part in a trillion of the carbon on or near the Earth's surface. It decays into nitrogen-14 with a half-life of 5,730 years. The activity in 0.035 ounces (1 gm) of carbon-14 is 14 disintegrations per minute. The average, 154-pound (70-kg) human is 18 percent carbon, or contains 28 pounds (13 kg) of carbon. Of that, only 4.8×10^{-7} ounces (1.36×10^{-5} gm) is carbon-14. Because of this very small amount, carbon-14 does not contribute greatly to the internal radioactivity of the human body. The β– decays from this radionuclide contribute about one millirem (0.01 mSv) per year to a person's total radiation dose. As a comparison, the internal exposure from potassium-40, expressed as a bio-adjusted dose, is 39 millirems (0.39 mSv) per year.

advantageous mutations remain in the population longer. This was used to explain an observation that the average genetic fitness of a population depends only on the rate of mutations and not on the degree of harm caused by each mutation. A central question thus formed: Is biological evolution speeded up by external, radiation-induced mutation, or is it inhibited? As the genetic mutation process is further studied, the debate continues.

The Nuclear Regulatory Commission (NRC) occupational eye lens exposure limit for nuclear industry employees is 1.5 rems (150 mSv) per year, and this is less than the 1.75 rems (175 mSv) that one can absorb standing on the beach in Guarapari, Brazil, for a year where the monazite sands are an important source of thorium. The highest recorded natural background radiation in the world is in the small town of Ramsar, in the Mazandaran Province of Iran, on the coast of the Caspian Sea. The hot springs in Ramsar bubble out radon gas from the interior of Earth, giving each person a dose of 2.6 rems (260 mSv) per year. Two years of living in Ramsar gives a person about the same radiation dose absorbed by a citizen of Hiroshima, Japan, surviving the atomic bomb blast that destroyed the entire city in 1945.

In preliminary studies, no obvious genetic mutations or abnormalities have been observed in residents of Ramsar, with the possible exception of a radioadaptive response of their lymphocytes. People in Ramsar show fewer induced chromosome aberrations compared with people in a nearby control area when exposed to 150 rads (1.5 Gy) of gamma rays. Constant exposure to radiation seems to make them less sensitive to tissue damage from higher radiation levels.

URANIUM AND PHOSPHATE MINING OPERATIONS

The heavy, black uranium-oxide ore, pitchblende, was known to miners well before the discovery of uranium as an element. In the year 79 C.E., uranium was used to give ceramic glaze a yellow tint and in making expensive Roman stemware. In 1789, Martin Heinrich Klaproth (1743–1817), a German chemist, discovered a metal with new and unique properties in a sample of pitchblende by a series of mistaken assumptions. He named the new element uranium, after the newly discovered planet Uranus.

High-grade uranium ore was mined in Colorado, in England, and in what is now the Czech Republic in the 19th century, but never in great quantities as its use was confined to pottery coloration. After radium was discovered in uranium ore in 1898 by Marie Skłodowska-Curie (1867–1934) and Pierre Curie (1859–1906), an industry was created, using radium for everything from illuminated watch dials to medical implants. The Shinkolobwe uranium deposit was discovered in Katanga, in what is now the Shaba Province, Zaire, in 1913—it was exploited by the Belgian company Union Minière duw Haut Katanga. Another rich lode was discovered near Great Bear Lake, Canada, in 1931, at a place named Port Radium.

The true importance of this element was not realized until the end of World War II, when the ability to react purified uranium in a self-propagating, energy-producing chain reaction was revealed. In 1956, the U.S. Atomic Energy Commission (AEC) set the price per ton for uranium ore artificially high, and an exploration and mining boom lasted until 1960. Uranium mines opened in New Mexico, Wyoming, Colorado, Utah, Texas, Arizona, Florida, Washington, and South Dakota, usually as open-pit or shaft mines. The general collapse of uranium prices in 1992 closed every pit and shaft uranium mine in the United States.

A small amount of uranium, about 4 percent of the worldwide uranium production, is still mined in the United States using in-situ leaching. Natural groundwater is pumped from a deep well, fortified with carbon dioxide and oxygen, and pumped back into the ground. Uranium oxide underground dissolves in the water and tends to flow back into the well, where it is picked up by the pump. Uranium is extracted from the water, which is sent back down for further leaching. The process is inexpensive, clean, and does not expose workers to radiation.

A difficulty in using conventional mining techniques for minerals in general is that digging a mine shaft invariably releases radon gas. A uranium or vanadium mine is particularly rich in radon. Radon remains confined to the ore and surrounding rock, displacing other gases, until it is disturbed by mining tools, at which time it expands into the confines of a mine shaft, tunnel, or the bottom of a pit. Workers breathe it. The decaying uranium and thorium are unusually concentrated in a mine, and the circumstances are ideal for a maximum radon exposure. Dust masks, used to prevent mining dust from invading the lungs, do not exclude radon, and heavy, bulky air tanks and respirators are necessary to prevent miners from being dosed with radon.

Currently, 23 percent of the mined uranium comes from Canada, 21 percent from Australia, and 16 percent from Kazakhstan. The remaining 40 percent of active uranium mining is distributed among South Africa, China, Ukraine, United States, Uzbekistan, Namibia, and

Uranium Mining Areas

- Texas Gulf Coast, TX
- Grant's Mineral Belt, AZ/NM
- Northern Arizona, AZ
- Marysvale, UT
- Paradox Basin, UT/CO
- Tallahassee Creek, CO
- Front Range, CO
- Northwest Nebraska, NE
- Washaki Sand Wash Basin, WY/CO
- Central Wyoming Basin, WY
- Wind River Basin, WY
- Powder River Basin, WY/MT
- Spokane, WA

© Infobase Learning

Uranium mining regions in the continental United States

An active open-pit uranium mine in Australia *(John Carnemolla, 2010, used under license from Shutterstock, Inc.)*

Niger. Nearly one-quarter of the world's known reserves of uranium are in Australia, which has no nuclear power industry nor any plans for one. Uranium is currently mined for export in the Kakadu National Park, in the Northern Territory, bringing about A$2.1 billion in the first five years of export, beginning in 2000. Unfortunately, the Ranger mine in Kakadu is atop an Aboriginal sacred religious site, and there is continuing controversy concerning possible harm to the environment in this area.

Most of the known uranium on Earth is actually dissolved in the oceans. The concentration is low, about 3.3 parts per million, but the ease with which seawater can be processed makes it almost a source of nuclear fuel. There is no problem of radon-222 release, as there is no mine tunnel or rock crevice for it to collect in and concentrate. All the radon in seawater is constantly emptied into the atmosphere, where it is severely diluted. Research into filtering uranium oxide out of seawater has been ongoing since the 1960s.

ALTITUDE-DEPENDENT COSMIC RAY EXPOSURE

Cosmic rays are extremely high-speed particles that collide with air molecules in the upper atmosphere, causing showers of ionizing radiation and exotic particles to rain down on Earth. Particularly intense bursts of cosmic rays are caused by solar energetic particle events, also called solar cosmic ray events. These solar flares are sporadic and do not seem predictable, although there is an 11-year cycle for the solar flare season. A solar flare can last from two minutes to several hours, producing vivid aurora effects and disruptions of wireless communications.

Fortunately for living things, most of the dangerous radiation is deflected by the geologic magnetic field of the Earth, and the atmosphere above provides very effective shielding. On the ground at sea level, there are more than 14 pounds (6 kg) of air above each square inch (6.5 cm²) of area. Air may seem a thin shield, but with the entire atmosphere overhead, the protection is substantial. It is the equivalent of having 32 feet (10 m) of water between outer space and the surface of Earth. Near sea level, the total yearly dose from cosmic rays is about 24 millirems (0.25 mSv). However, standing in the eastern Rocky Mountains for a year, one would receive 63 millirems (0.63 mSv). Cosmic ray exposure is altitude dependent. The higher up in the atmosphere a person is, the more radiation is absorbed.

This phenomenon gives airline pilots and flight attendants the greatest radiation dosage of any profession, far above the dose absorbed by most nuclear plant workers. An airline flight of two hours at an altitude of 30,000 feet (9,000 m) will approximately double one's background radiation dose for the day. A solar energetic particle event could increase this dose by a factor of 10.

At 30,000 feet (9,000 m) above ground, the typical altitude of a jet airliner flight, at solar minimum and 35 degrees North latitude, the dose is 325 microrems (3.25 μSv) per hour, and at 75 degrees North the dose is 406 microrems (4.06 μSv) per hour. At solar maximum, the numbers drop to 285 microrems (2.85 μSv) and 324 microrems (3.24 μSv) per hour, respectively. At the extremely high altitude of 80,000 feet (24,000 m), usually attainable only by military aircraft, the dose rate tops out at 2.06 milli-rems (20.6 µSv) per hour at 70 degrees North latitude and solar minimum.

The cancer risk factors for background radiation dosage have been estimated by the National Council on Radiation Protection and Measurement. The cancer probability for a population of all ages is 0.0005 per rem (0.05 per Sv) of exposure. If, for example, a population is exposed to 20 millirem (200 µSv) of ionizing radiation, then one cancer caused by radiation might appear in a population of 100,000. For infants, the risk is approximately twice that. To put this in perspective, the average cancer rate in 100,000 people is one in three, or about 30,000 people will contract cancer in their lifetimes regardless of their radiation exposure.

RADIATION IN FOOD

There are two ways to consume radiation in food. The most common ingestion of a radioisotope is by eating food that contains potassium with its very small percentage of the nuclide potassium-40. All organic food also contains carbon-14, but in concentrations small enough to be detectable only with very sensitive radiation detection instruments. Potassium concentrates in certain foods, such as orange juice, potatoes, bananas, avocados, tomatoes, broccoli, soybeans, brown rice, garlic, and apricots. It is possible for trace amounts of uranium or thorium to appear in food, leached from the topsoil.

Radiation also turns up in food, not by biological process, but by radioisotopes being dropped on it out of the sky, contaminating the surface of plants. This was a problem in the 1950s and '60s, as aboveground nuclear weapons tests were throwing tons of fission product–contaminated dust into the upper atmosphere. This dust would eventually come down in rainwater, distributing a thin layer of radioactive substances wherever rain would fall. Fruits and vegetables eaten raw were particularly prone to have radioactive contamination, and washing fruits, tomatoes, and lettuce did not completely eliminate the problem. The fission product–contaminated dust was termed *fallout*. New fallen snow, which did not drain away and disperse quickly, as would rainwater, seemed to emphasize the problem, and the heightened activity level could be detected with a handheld *Geiger counter*.

In addition, there is a secondary effect of this surface contamination. Grass is obviously coated with a thin layer of rain-borne fallout, and cows, sheep, and goats eat grass. Grass is given a detailed chemical analysis

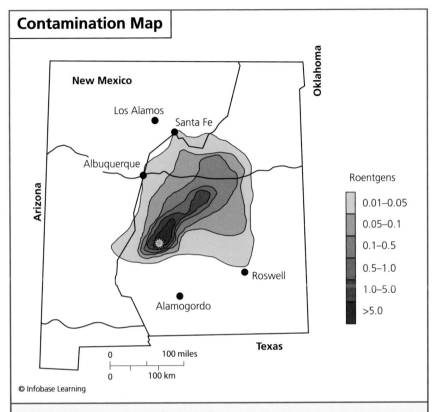

Contamination Map

New Mexico

Los Alamos
Santa Fe

Albuquerque

Arizona

Oklahoma

Roentgens

	0.01–0.05
	0.05–0.1
	0.1–0.5
	0.5–1.0
	1.0–5.0
	>5.0

Roswell

Alamogordo

Texas

0   100 miles

0   100 km

© Infobase Learning

A map showing the spread of radioactive dust in the United States from the first aboveground nuclear weapons test near Alamogordo, New Mexico, on July 16, 1945; radiation areas were plotted from ground-level readings of radiation taken shortly after the test firing using Geiger-Müller counters. The wind carried fission debris and unburned plutonium northeast, and fallout was detectable but not at a dangerous level as far away as Santa Fe and Albuquerque.

in a cow's digestive system, with various nutrients sorted by elemental composition and biological need. Iodine in the grass is divided out and diverted to milk production, where it is supposed to feed growing infant cattle. A particularly dangerous fission product, made in nuclear weapon detonations or in nuclear power production, is iodine-131. Chemically, it is the same as natural iodine-127, which is stable and nonradioactive. The iodine-131 on the surface of the grass is digested, put in milk, and extracted from the cow. The half-life of iodine-131 is only eight days, but the turnaround time for dairy products is fast. Little of the iodine-131 is wasted, and it turns up in milk, ice cream, cheese, yogurt, sour cream, and any food that uses a dairy product. The human digestive system further parses out the iodine-131, putting it directly into the thyroid gland, where its rapid beta decay causes much damage.

Public outcry and scientific concern over this worldwide contamination of food caused the drafting of the Treaty Banning Nuclear Weapon Tests in the Atmosphere, in Outer Space and Under Water (Limited Test Ban Treaty). The treaty was signed on August 5, 1963, by Dean Rusk of the United States, Andrei Gromyko of the Soviet Union, and Sir Alec Douglas-Home of the United Kingdom. The treaty went into effect on October 10, 1963, and the rapid loading of the atmosphere with radioactive dust essentially stopped. Occasional nuclear weapons tests by countries that did not sign the agreement, such as China, have never approached the intensity of testing in the early 1960s, when multiple aboveground weapons could be set off in a single day. An atmospheric monitoring station in Wellington, New Zealand, detecting the concentration of radioactive carbon-14 in the air, had recorded a doubling of the activity between January 1955 and October 1963. After the test ban, the atmospheric activity began a long drop, and by 2009 it had fallen to 1955 levels.

SITE CONTAMINATION

Beginning with World War II in 1941, hundreds of nuclear research, material processing, and power production sites were built in the United States. For 50 years, federally and privately funded properties were devoted to nuclear work in all 50 states. Each site, from nuclear fuel processing plants to atomic bomb test areas, involved radioactive materials, and to some extent all were contaminated with radioactive materials. In some locations involved in military testing, such as Jackass Flats, Nevada, no attempt was made to contain or control fission product or fissile fuel release, but by the 1960s most industrial plants and test areas went to great lengths to avoid contamination of the surrounding area. Still, wherever radioactive materials were used in industry or research, some of it was bound to wind up on the ground. As nuclear activities spread around the globe, so did industrial sites, followed closely by radioactive contamination, in hundreds of locations outside the United States.

In the most extreme cases of site contamination, entire islands in the South Pacific Ocean were either wiped off the map or rendered uninhabitable by aboveground weapons testing by the United States. Other countries, such as Great Britain, France, and the Soviet Union, conducted aboveground tests during the same 15-year period and managed to contaminate Kazakhstan, Algeria, and Australia. By now, 50 years later, most of the *fission products* left behind on the surface have decayed to safe levels, but the signature radiations left by these tests will remain detectable for hundreds of years, clearly marking the test sites.

The next chapter discusses details of industrial radiation sources in common use. At the current level of technical sophistication in the industrial world, radiation seems to be everywhere, even in countries where nuclear power and weapons are banned.

4 Industrial and Medical Uses of Radiation

Within a short time, perhaps weeks, after the formal discovery of electrically produced radiation in 1895 by Wilhelm Roentgen (1845–1923), it was put to both medical and industrial use. In fact, as early as 1881, Pulyui lamps were mass-produced in the Ukraine for use in X-ray analysis of broken bones, before the X-ray was correctly described and interpreted by Roentgen. In the late 19th century, radioactivity in minerals was discovered. Chemists and physicists, often seeking practical applications of novel discoveries as sources of research funding, quickly found profitable uses for radiation in industry as well as medicine. At the turn of the century, radiation began to turn up in everyday life, occasionally with disastrous results, and it would take 50 years before the impact and necessary cautions for radiation use were fully realized.

Today, radiation is in daily use in all aspects of industry and medicine. The obvious and most intense use is in nuclear power, in which radiation release in a tightly controlled space is a primary source of power, but there are other, less obvious uses in everything from seafood preservation to airport security. This chapter introduces the reader to some of the current uses of radiation in the industrial world and to the importance of careful control of radioactivity sources. There even needs to be control over discarded radiation sources. Awareness of these issues will become increasingly important as the level of background radiation is artificially incremented in the next 100 years.

SMOKE DETECTORS, SELF-LUMINOSITY, AND LANTERN MANTLES

About 93 percent of all homes in the United States have at least one smoke detector installed on a ceiling, and at least half of these units are ionization detectors. An ionization detector uses an electric voltage established between two electrodes hanging in the air. A speck of radioactive material near the electrodes causes the air to ionize, and ionized air conducts electricity, so a steady current of electrons flows between the electrodes. If a few particles of smoke wander into the space between the electrodes, the current drops, as the smoke particles absorb the ionizing radiation and degrade the ionization. Burn a slice of bread in the toaster-oven, the current drops, and the detector switches on a loud, screeching horn.

This ionization suppression by small quantities of smoke was discovered accidentally by Swiss physicist Walter Jaeger in the late 1930s. Jaeger was trying to develop a sensor for poison gas, and he predicted that gas molecules flowing between the electrodes would bind to the ionized air molecules and alter the current flow. He induced the air ionization using a small, radioactive alpha particle source. He powered up the equipment, introduced a controlled leak of poison gas across the electrodes, and nothing happened. Neither the degree of ionization nor the voltage across the electrodes could be adjusted to make the device work as predicted. Becoming frustrated, Jaeger lit a cigarette and took a puff. The experimental apparatus reacted to the cigarette smoke instantly.

The first ionization smoke detectors, using radium-226 as an alpha particle source, went on sale in 1941. They were expensive and were sold almost exclusively to hotels, where they were used sparingly. Home smoke detectors were not available until the 1960s, and they were still expensive. The alpha particle source was changed to a speck of polonium-210, which was less costly than the radium-226, but it had a half-life of only 138 days. After a year of operation, the detector's alpha source was too weak to function.

A more practical unit was manufactured by the Statitrol Corporation beginning in 1975. The short-lived polonium-210 source was replaced by 0.0053 ounces (150 milligrams) of americium-241, giving about 0.9 microcuries (33.3 kBq) of radiation with a 432.2-year half-life. The alpha particle source of americium-241 would outlive the device, and the batteries were also simplified to a readily available nine-volt unit. Operation of this improved ionization detector involves alpha radiation at more than five MeV per unit, plus accompanying low-level gamma rays. These sources

Internal components of an ionization smoke alarm; the radioactive part has a yellow and magenta radiation warning sign stuck to it. *(Andrew Lambert Photography/Photo Researchers, Inc.)*

are physically small, and it would seem important to keep them out of the food supply. The theoretical danger of alpha sources scattered into the environment remains for hundreds of years, although as a practical measure the Nuclear Regulatory Commission (NRC) does not consider this a radiation hazard. There is no organized program for controlled recycling of discarded units.

Self-luminosity is the ability of an object to light itself so as to be visible in the dark, without the use of electricity, a flame, or any moving parts. The principle of self-luminosity was discovered by Marie and Pierre Curie, who found that when mixed with a fluorescent material, such as zinc sulfide, any compound of radium will cause visible light. Marie Curie kept a small vial of a radium salt with zinc sulfide dissolved in water in her smock to show visitors to her laboratory. While it did make a pretty blue glow, it was dangerous to carry around. By 1903, the Curies were using dishes of the glowing substance around the garden as mood lighting for parties, but both were starting to suffer from the effects of radiation from the radium.

In 1917, the U.S. Radium Corporation began operation in Orange, New Jersey, to make Undark, the trade name for luminous paint made with radium-226 with its 4.8 MeV alpha particle and its 1,600-year half-life. This product gained a large market, making watch and alarm clock dials visible in the dark, and the company processed about 1,000 pounds (454 kg) of carnotite ore daily to extract the radium. Unfortunately, a large percentage of workers in the factory, primarily women painting watch dials, became contaminated with radium through ingestion. The problem reached epidemic proportions as employees began to die of radium jaw, an ailment confined to radium dial painters. Litigation and unfavorable publicity caused the company to close down in 1927.

The demand for glow-in-the-dark items continued to grow as U.S. Radium went out of business, and other companies were formed to meet the demand. Radium dials were considered of strategic importance to the military, and during World War II millions of glowing items were made, including all aircraft instruments, military watches, compasses, and even airfield landing lights. Theaters and public buildings were equipped with radium-painted exit signs. They were too convenient not to use for public safety, even given the danger to the manufacturing workers. It was reasonably safe to stand under a radioactive exit sign, as long as one was not tempted to eat it, but to be closely exposed to thousands of signs per hour as they were being built was a definite hazard.

In 1983, the abandoned U.S. Radium Corporation plant in Orange, New Jersey, was declared a Superfund site by the U.S. Environmental Protection Agency (EPA), and from 1997 through 2005, 1,600 tons (1,452 mt) of radium-contaminated material was excavated and buried off-site as radioactive waste. There are an estimated 250 other commercial and residential properties requiring EPA action due to radium contamination.

Up until the commercial success of electrical incandescent lighting, about 1910, most lighting in the world was accomplished with burning oil or gas. This was a very inefficient, dim way to produce illumination, and the light was colored yellow. Most of the energy was wasted as infrared light, below the frequency threshold of human eyes. Gas and kerosene lights, from streetlights to bedside night-lights, were vastly improved in 1885. Carl Auer, Freiherr [Baron] von Welsbach (1858–1929), an Austrian scientist and inventor, developed a method of transforming the heat from a gas flame to white light, with a minimum of infrared waste. He patented the gas mantle, or auerlicht, consisting of a thin cotton sock soaked in a mixture of magnesium oxide, lanthanum oxide, and yttrium oxide. When heated by a flame, the cotton burns away and leaves a fragile gauze of ash. Auer's original mantle glowed green and as a commercial item did not sell well.

In 1890, Auer introduced a new mantle formula, based on a mixture of 99 percent thorium dioxide and 1 percent cerium oxide. The thorium glowed a brilliant white, and the popular new product spread quickly through Europe and the United States. Gaslight mantles were used widely, well into the 20th century, with the application eventually displaced by electric lighting. The thorium mantle is still in limited use, for example, in the Coleman white gas camping lantern. The thorium is radioactive, but only slightly. Working under a thorium mantle lantern every weekend for a year leads to a 0.3 to 0.6 millirem (0.003 to 0.006 mSv) radiation dose. Ingestion of an entire mantle would lead to a more significant 200 millirem (2 mSv) accumulated dose.

INDUSTRIAL RADIOGRAPHY

For the past century, radiography has been used extensively for medical diagnosis. X-rays, produced electrically, are aimed through flesh for a short exposure time, and varying densities of tissues tend to block the rays from coming through on the other side. The X-rays then expose a large sheet of photographic film, held in a light-tight covering that is incapable of stopping the radiation. When the film is developed in chemical baths and dried, it shows an image of the internal structures through which the X-rays traveled, revealing details, such as the exact positions of broken bones, that were not obvious from outside the body. There are variations, such as the very small, in-the-mouth films used by dental X-ray technicians, and even digital imaging methods that no longer use photographic film, but the concept of the medical X-ray remains simple. A point source of X-rays on one side of a body exposes sensitive photo material on the other side of a body, showing what is inside. The technique is *radiography*. A lesser known form of radiography is used in many industries worldwide.

Industrial radiography followed closely behind medical radiography from its inception in 1895. It was found valuable in inspecting critical metal components, such as boilers, tanks, pipes, and valves, for dangerous defects. A pipe carrying high-pressure steam could look perfect on the

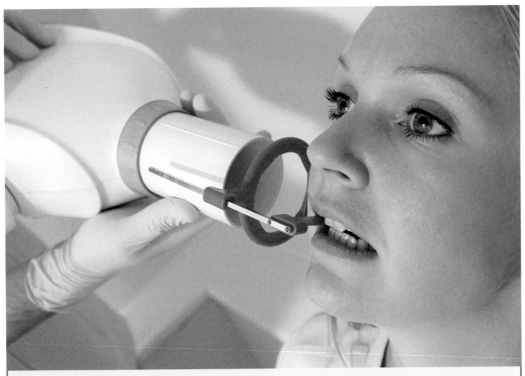

A woman having a dental X-ray *(Monkey Business Images, 2010, used under license from Shutterstock, Inc.)*

outside, and even perfect on the inside, but a small crack or void in the metal of the pipe wall could result in an unexpected steam explosion. Just as a medical X-ray could reveal tuberculosis or breast cancer, radiation shot through a metal object could show otherwise invisible defects.

An immediate problem with industrial radiography was that medical X-rays were too weak to penetrate the steel in a steam pipe. Radiation further up the spectrum, with a higher frequency and more energy per photon, was needed. Gamma rays from available minerals were put to use in nondestructive testing of steam equipment and ship hulls in the early days of radiography. Before World War II, the only radioactive isotopes obtainable were radium-226 and radon-222. These were alpha emitters, and alpha particles have no penetrating power, but accompanying the alphas were moderately strong gamma rays. These materials were very expensive in useful quantities, and they were dangerous to use.

Cobalt-60 was found to be a particularly good material, having a 5.27-year half-life and a 1.3 MeV gamma ray. Cobalt-60 is a fission product, available in nuclear power waste, and it can also be made from natural, nonradioactive cobalt-59 by exposing it to neutrons in a reactor. In the 1950s, radiography using fixed gamma ray sources became an essential part of nondestructive testing, particularly for grading welds on pipes carrying high-pressure steam. The technique quickly found other vital uses, testing steel or aluminum pressure vessels, high-capacity storage containers, pipelines, and structural welds. It could be used to find rebar steel or conduits in flat concrete. By the 1960s, radiography was being used to see the inside of machined parts.

The technique is mechanically simple and requires no electrical hookup. It can be used in remote situations, such as pipelines running through the wilderness. The radioactive source can be cesium-137 or cobalt-60, but the most common source material is now iridium-192m, a metastable gamma emitter with a half-life of only 74.3 days. It is held safely in a solid block of metal, usually lead, called the camera or the torch. A few inches of lead can block even an energetic gamma ray source and make it safe to handle. Sheets of photographic film, shielded from light contamination, are taped to the outside of a pipe, over the weld to be tested and covering the entire circumference. The front of the lead gamma ray camera is hinged, and when opened a wide spray of invisible, noiseless gamma rays emit from the source. The opened camera is placed at the center of a pipe, near the welded seam, and a short section of the pipe is flooded with gamma rays. After a few minutes of exposure, the films are removed and sent to the darkroom for processing. The camera is removed from the pipe, and the shield door is closed. The developed films are then closely studied on a light table, with white light back illumination, and any anomaly or imperfection in the weld will show up on the X-ray. The films are then stored in an index, which refers to the pipe and the weld by number, for any future reference. Everything from power plants to fuel refineries is built using this test procedure, and it ensures sound welds and safe pipe runs.

With growing security concerns for cargo entering the country, means for radiological imaging of entire tractor trailer truckloads and intermodal containers are currently being developed. Sending radiation completely through a truck requires high energies and quantities of radiation, and small, solid-state sources cannot supply enough gamma rays for these scans. High-energy betatron X-ray sources are being tried, as well as muon sources. Muons have a superior penetrating capability over X-rays or gamma rays, but they are difficult to produce, requiring high-energy cosmic ray disintegrations or the equivalent events in large particle accelerators. These large-scale systems are currently under research.

Airport security is a current example of industrial radiography. Dual-energy X-rays are used for baggage screening, looking for weapons, bombs, or bomb materials in carry-on luggage as air passengers move to the gates. Each bag is moved by conveyer belt into a shielded box, where it is stopped and subjected to a low-energy and then a high-energy X-ray illumination.

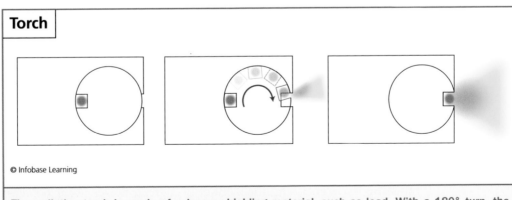

Torch

© Infobase Learning

The radiation torch is made of a heavy shielding material, such as lead. With a 180° turn, the turntable moves the radioactive source from a fully safe position inside the shield to full exposure, with a clear path opened to outside the shield.

The two images are collected in turn by a solid-state, electronic imaging device, sensitive to X-rays. No chemical development of the images is necessary, as they are digitized and processed by a computer. The two images, low and high X-ray, are combined on the computer view-screen, assigned the colors orange and blue. The resulting image shows low-density materials, such as paper or cloth, as orange, and high-density objects, such as the steel parts of a pistol, as blue. A great deal of interpretation by human operators is necessary for this system to do the job of screening potentially troublesome cargo.

A significant danger to using portable isotope radiation sources for industrial radiography is the accidental loss of these sources. If a gamma-ray source finds its way into the hands of someone who has no idea what he is holding, the results can be and have been disastrous. Cobalt-60 sources have been discarded or misplaced in Russia, Brazil, Mexico, and Morocco, with deadly consequences. The fundamental problem, that radiation cannot be sensed, is particularly acute when these materials slip the controls, and tight regulation is absolutely necessary.

FOOD IRRADIATION

An early application of the newly available gamma-ray isotopes after World War II was the irradiation of food. Exposure to ionizing radiation can be beneficial to food, while causing it no harm. A sufficient dose of gamma rays destroys microorganisms, bacteria, viruses, insects, insect larvae, and insect eggs. It does so without any insecticides or antibiotics, introducing no chemicals to the food, and delays ripening and sprouting. It increases juice yield in fruits and improves rehydration of dehydrated foods.

The effect is similar to the pasteurization of milk by heat, only it does not involve heating the food or partially cooking it. Irradiation of food is currently allowed in more than 40 countries, and an estimated 551,000 tons (500,000 mt) per year of food is irradiated worldwide. Irradiated food sold in the United States must be labeled "Treated with irradiation" or "Treated by radiation" and must be identified by the *radura* logo at the point of sale. The radura is a green circle, pierced with five radiating lines, enclosing the symbol of a green flower. After 60 years of irradiated food, consumers are still leery of the concept, and currently the only foods processed by gamma rays are fresh tropical fruit from Hawaii or Florida, dehydrated spices, and ground meat products.

Certain supermarkets specializing in "organic" foods prefer not to carry irradiated items.

Cobalt-60 is the radiation source of choice for food processing. Treatment consists of an automated line of food pallets, which are run over a cobalt radiation source on a conveyor belt and stopped for the irradiation period, ranging from a few minutes to hours, depending on the dosage required. The cost of this treatment can range from one cent per pound for deinfestation of fruit to as much as 20 cents per pound for the destruction of *E. coli* in ground meat. Workers at irradiation plants are separated from the process by heavy concrete walls.

The internationally recognized radura logo, used to label irradiated food *(AP Images)*

Applications of food irradiation are roughly divided into low, medium, and high gamma radiation doses.

The low food dose is up to 100 kilorads (one kGy), which would be quite high if food were alive. At three to 15 kilorads (0.03 to 0.15 kGy), sprouting is inhibited in bulbs and tubers. Fruit ripening is delayed by 25 to 75 kilorads (0.25 to 0.75 kGy). Food-borne parasites and general insect populations are eliminated by seven to 100 kilorads (0.07 to 1.00 kGy).

The medium dose is considered to be between 100 and 1,000 kilorads (one to 10 kGy). Microbial spoilage under refrigeration is reduced by a gamma-ray exposure of 150 to 300 kilorads (1.50 to 3.00 kGy), and this prolongs the shelf life of meat, poultry, and seafood. Pathogenic microbes are reduced in the same fresh and frozen items by 300 to 700 kilorads (3.00 to 7.00 kGy). The hygienic quality of spices is improved by reducing the number of included microorganisms by 1,000 kilorads (10.0 kGy). The infestation is reduced by several orders of magnitude, and treated spices do not pass along microorganisms to final products.

The high dose rate, which is above 1,000 kilorads (10 kGy), is not currently approved for use in the United States by the Food and Drug

Irradiated strawberries on the left, and normal strawberries on the right *(Cordelia Molloy/Photo Researchers, Inc.)*

Administration (FDA), but it has been studied extensively for several decades. Packaged meat, poultry, and their products can be sterilized with 2.5 to 7.0 megarads (25.0 to 70.0 kGy). Food subjected to this dose is shelf stable without refrigeration. Hospital diets, prepared for people who have no natural defense against microorganisms, are also subjected to 2.5 to 7.0 megarads (25.0 to 70.0 kGy). This dose also yields an increased juice yield from picked fruit and improves the rehydration of dehydrated items. NASA is authorized to irradiate frozen meat for astronauts at 4.4 megarads (44.0 kGy).

MEDICAL STERILIZATION

Every instrument and device used in a hospital for the treatment of patients, from the smallest needle to robotic surgery arms, must be sterilized of all microbiotic life. Any surgery or breaching of the skin on a human being is an opportunity to introduce germs into the system, and

THORIUM TOOTHPASTE

Carl Auer, Freiherr [Baron] von Welsbach, the man who invented the gas lantern mantle, also started a company in Berlin. Auergesellschaft was an industrial firm founded in 1892, which constructed a factory in 1926 in Oranienburg, 15 miles (24 km) northeast of Berlin. The plant processed uranium ore, producing industrial quantities of thorium and radium. The waste product, uranium, was dumped into a growing pile behind the plant.

Always looking for consumer products to use his thorium and radium, Auer came up with a new product, Doramad radioactive toothpaste, containing thorium and radium. The advertisements made it very attractive to buyers concerned about their oral hygiene:

Use toothpaste with thorium! Have sparkling, brilliant teeth—radioactive brilliance!

A full explanation was printed on the back of the tube:

What does Doramad do? Its radioactive radiation increases the defenses of teeth and gums. The cells are loaded with new life energy, the destroying effect of bacteria is hindered. This explains the excellent prophylaxis and healing process with gingival diseases. It gently polishes the dental enamel so it turns white and shiny. Prevents dental calculus. Wonderful lather and a new, pleasant, mild and refreshing taste. Can be applied sparingly.

If the product had enough radiation to hinder the effect of bacteria, then it was much too dangerous to have in the mouth. Users began to complain of tooth loss and dissolving jaws, and production of the product was quietly cancelled. It remained on some store shelves until the end of World War II. The substantial waste pile of uranium behind the factory building was used in an unsuccessful attempt by the Heereswaffenamt, the German army weapons agency charged with research and development of weapons, to build a nuclear reactor as the war raged in Europe.

In fall 1944, the advancing Allied Expeditionary Force, pushing the German army from France, discovered that Auer had taken over the French uranium-processing company Terres-Rares during the German occupation. Auer had shipped a massive supply of processed thorium back to Germany. This concerned the scientific exploratory mission, Operation Alsos, led by Colonel Boris T. Pash (1900–95). Scientists in Pash's

(continues)

(continued) _____

group feared that Germany was working on an advanced, thorium-fueled nuclear reactor, which would be ahead of the top-secret effort under back in the United States.

Further probing by Operation Alsos revealed that Auer was simply trying to corner the thorium market in anticipation of a postwar craze for radioactive cosmetics. Unfortunately for Auer, the demand failed to materialize. In 1958, Auergesellschaft merged with the Mine Safety Appliances Corporation, a multinational U.S. corporation.

Operation Alsos personnel were similarly disappointed with all their subsequent findings concerning the German atomic bomb project. Finally, as the war with Germany came to a close in 1945, the American scientists found the last refuge. It was an experimental nuclear reactor, built into the floor of a beer-cellar under the castle in Haigerloch, high above the Eyach River. The experiment had proven unsuccessful.

for medical procedures to improve the condition of a patient, this must be avoided. Steam is used for sterilization, as well as many chemicals, such as ozone and bleach. A practical, nondestructive way to kill organisms living on medical equipment is radiation.

Electron beams, X-rays, gamma rays, and accelerated subatomic particles are used for medical sterilization. The most common method is to use gamma rays from a fixed cobalt-60 source. The problems with using cobalt-60 as a gamma source are common to all industrial uses of radioactive isotopes. The gamma rays cannot be switched off, and the source remains in full operation at all times, even when the sterilization equipment has outlived its usefulness and has been discarded. It is important that a gamma-ray sterilizer be accounted for even when it is junked. Bulky, heavy shielding must be used to protect the equipment operators.

Disposable medical equipment, such as syringes, needles, cannulas, and intravenous sets, are all sterilized by gamma rays. An alternative is the electron beam process, which provides an even higher dose rate than cobalt-60. It has the advantage of being able to be turned on and off. At the end of its useful life, the electron beam projector can simply be thrown away, and it will have accumulated no radiation. The higher dose rate means that items do not require as much time under radiation, and polymer plastics being irradiated are subjected to less degradation. While turned on, the electron beam is every bit as dangerous as gamma rays, and much shielding is necessary. The equipment is more expensive than a cobalt-60 facility and is subject to failure, just as is any complex electronic device.

An interesting nonmedical use of radiation biocide is on the mail delivered to government offices in Washington, D.C., which is sterilized before delivery using cobalt-60 and electron-beam accelerators. This extreme precaution is specifically used to kill anthrax bacteria that may be sent through the mail to political offices in the nation's capital. This expensive action began in November 2001, after individuals in the government and the news media were targeted by an apparent terrorist tactic using the mail service.

RADIOPHARMACEUTICALS

Many medicines containing radioactive nuclides, *radiopharmaceuticals,* are used for diagnosing and treating human illnesses. There are 28 nuclides, or atomic species, presently in use. There are 31 medical tests performed using just one of these nuclides, technetium-99m. The radiopharmaceuticals are manufactured by neutron activation of stable nuclides using nuclear reactors or extracted chemically from fission by-products of nuclear-power waste.

Most uses for radiopharmaceuticals are in diagnosis, and there are two modes of use, imaging and non-imaging. In imaging diagnosis, the positions of a radionuclide chemical administered to the body reveal by scanning, using a narrow-beam radiation detector. Results of the scanning are then assembled on a computer monitor, showing an image of where in the body the radioactivity has collected. For example, technetium-99m can be injected into the bloodstream. An image of its position and concentration around the heart muscles after vigorous exercise can then indicate clogged or open arteries in the heart.

There are also therapeutic uses for radionuclides. Yttrium-90 compounded with silicon, for example, can be injected into the arteries to treat arthritis or inserted into malignant tumors to treat cancer.

DENTAL AND BODY X-RAYS

For a person not working in a radiation-prone profession, such as uranium mining, the highest radiation dose ever received will probably be a diagnostic medical X-ray. Almost everyone has had at least one dental

X-ray. An X-ray shot through a human body is not without risk, as it is an ionizing radiation, but the medical industry is very careful with diagnostic radiation and has gone to every effort to minimize the accumulated dose acquired in a dental or body X-ray. No more radiation is used than necessary to form an image in X-ray sensitive photographic film, a phosphor screen, or a direct semiconductor detector, similar to a digital camera.

The dose of radiation given by a diagnostic X-ray is officially posted in millirems (μSv), but to make this abstract unit meaningful to patients X-ray doses are often expressed in *background radiation equivalent time (BRET)*. BRET is expressed in days, months, or years and is the amount of time standing in the open that would be necessary to absorb an equivalent dose from natural background sources, on average.

A dental X-ray involves probably the least X-ray dosage. A series of four intraoral radiographs, involving small squares of photographic film held in the mouth, results in a dose of 3.8 millirems (38 μSv). The BRET is 4.8 days.

A woman being X-rayed with a fluoroscope *(Steve McAlister/The Image Bank/Getty Images)*

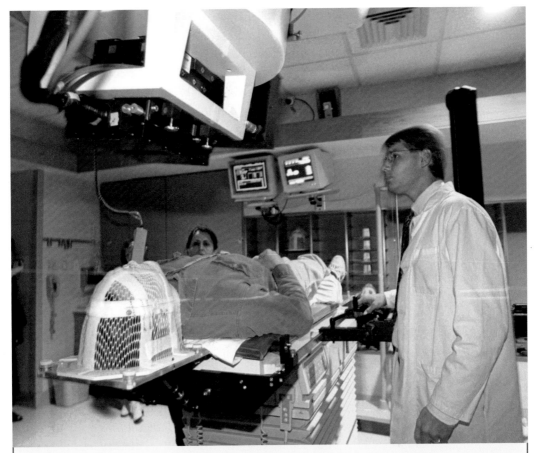

Radiation treatment for an inoperable brain tumor; in cases such as this, radiation doses can be administered in quantities over 10 rems (100 mSv). *(AP Images)*

A dental panoramic radiograph, covering the entire jaw in one 180-degree sweep, gives a 2.6 millirem (26 μSv) dose, with a BRET of 3.3 days.

The most severe medical X-ray dose normally administered is probably in the computed *tomography* (CT) scan. A CT scan of the entire body, the abdomen and pelvis, or simply the colon, results in an accumulated dose of one rem (10 mSv), or a BRET of three years. A CT of the brain gives less dosage, at 0.2 rem (two mSv), or a BRET of eight months, and a head CT for calcium scoring is 0.3 rem (three mSv), with a BRET of one year.

The least medical X-ray dosage is probably from a bone densitometry measurement, at 0.1 millirem (0.001 mSv). The BRET is less than one day, making it an even smaller dose than a dental X-ray. A mammogram exposes a woman to 70 millirems (0.7 mSv), or a BRET of three months. A common chest X-ray gives an exposure of 10 millirem (0.1 mSv), and a BRET of 10 days. For an infant, the most dosage given is for the void cystourethrogram, in which the urinary bladder is watched by X-ray while it is emptied. The dose is 80 millirems (0.8 mSv), with a BRET of three months, or the child's lifetime to this point. For a five- to 10-year-old child, the flesh is more dense, and a 160-millirem (1.6-mSv) dose is necessary for the same procedure. The BRET for a child is six months.

CAT SCANNING AND MAGNETIC RESONANCE IMAGING

A CAT, computed axial tomography, or CT scanner was first used in medical practice in London, United Kingdom, in September 1971, finding a cerebral cyst in a patient at Atkinson Morley Hospital in Wimbledon. The device, using an X-ray source mounted on a rotating ring opposite a scintillation detector, was invented by Sir Godfrey Newbold Hounsfield (1919–2004) and Allan MacLeod Cormack (1924–98). The two men shared the Nobel Prize in physiology or medicine in 1979 for having developed this valuable diagnostic technique.

A CT scan is a cross-sectional picture through a solid object, or person, showing the internal structure as a two-dimensional representation on a computer screen. The patient lies on a bench, and the scanner moves around in a 180-degree arc, stopping at one-degree intervals to take a sampling of the body's ability to stop X-rays, or its density. In the original machine, it took five minutes to take each density reading, and once data from all 180 angles through the body were collected, it took 2.5 hours on a mainframe computer to algebraically construct a tomogram, or the cross-sectional image. In honor of Sir Godfrey Hounsfield, the unit of measure of X-ray density used in CT scanning, the Hounsfield unit, or HU, is named for him.

An X-ray running through thin air is –1,000 HUs. Through water, or most solid flesh, X-rays encounter zero HU. Through bone is 1,000 HUs. Cranial bone is the most dense and can reach 4,000 HUs.

Nuclear magnetic resonance imaging (MRI) is not technically a medical diagnostic technique that involves ionizing radiation. It does, however, involve immersing the human body in electromagnetic radiation in the form of radio waves, and it deserves brief mention.

The first MRI image was published in 1973. As a competitor for the CT scan, an MRI scanner is generally superior for tumor detection and identification in the brain, while the CT is better for solid tumors in the abdomen and chest. The major advantage is that the MRI produces no accumulated radiation dose, so a patient can be given an MRI scan several times, whereas CT scans must be limited.

The MRI technique involves a property of protons. Protons will resonate in the presence of a radio wave of a specific frequency in a static magnetic field of a specific strength, and this resonance may be detected remotely. Protons stuck together in a nucleus have a lesser resonance tendency and at a different magnetism/radio frequency setting, but hydrogen has in its nucleus only one proton, and it produces a robust resonance

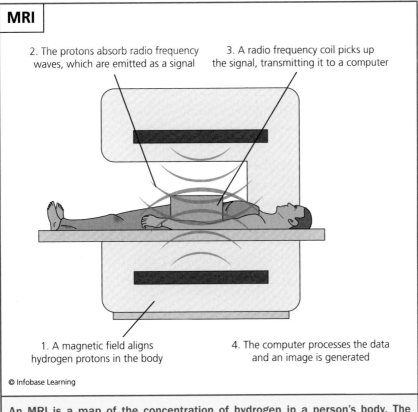

MRI

2. The protons absorb radio frequency waves, which are emitted as a signal

3. A radio frequency coil picks up the signal, transmitting it to a computer

1. A magnetic field aligns hydrogen protons in the body

4. The computer processes the data and an image is generated

© Infobase Learning

An MRI is a map of the concentration of hydrogen in a person's body. The hydrogen nuclei in a specific magnetic field strength react to radio waves of a specific frequency. The magnetic field through the body varies from top to bottom, and the vertical position of reacting nuclei can be mapped by noting the frequency at which they react. Hydrogen at the top of the body react at a higher frequency than those at the bottom.

signal. A human body is mostly water, with different tissues having varying concentrations of water. Bone, for example, has much less water content than does spinal fluid.

The MRI scanner maps the concentration of water in an axial slice of a body using a strong magnetic field and a radio signal of varying frequency. The magnetic field is produced in the body using a very strong electromagnet, typically with a maximum B-induction of 1.5×10^4 gauss (1.5 T). The magnetic field is graded, such that it falls off in a linear manner with distance. At a given magnetic field intensity, or B-induction, a single proton will resonate at only one specific frequency. By establishing a fixed B-induction at a point in the patient, the hydrogen at that point may be resonated by radio frequency radiation at a specific frequency. The strength of the resonance indicates the amount of water present at the spatial point. By sweeping the body in varying radio frequency radiation and biasing the magnetic field, a detailed image of the water content can be constructed with a computer. The heavy banging noise that seems to come from the MRI scanner is the sound of magnetic gradient coils turning on and off. The patient is affected in no detectable way by having the body hydrogen resonated, although the patient must remain perfectly still for meaningful images to be constructed.

These are the many types of radiation across a broad spectrum. Each type has properties and sources, natural and artificial. Each has its uses in the industrial world, and each has effects on people and objects. These radiations are generally invisible and are not obvious in daily commerce. How are these elusive forces measured, detected, or even found? There are many ways, and they are disclosed and reviewed in the next chapter.

5 Radiation Detection and Measuring

Ionizing radiation was a completely unknown feature of the natural world until it was first detected, probably by Johann Hittorf (1824–1914), a German physicist. Hittorf was investigating high-voltage vacuum tubes at the University of Münster in Germany in 1875. He noticed that his photographic plates, or film, were fogged by being anywhere near his apparatus, but he was not sure why. Something about the experimental equipment was ruining his pictures. Although he did not fully grasp the significance of his observation, Hittorf had discovered a property of photographic emulsion that is still in use for radiation detection, even as camera film is being rendered obsolete by digital imaging.

This chapter is an outline of the most important radiation-detection techniques developed in the past 100 years, all exploiting an effect of radiation on matter. The sidebar on pages 92–93 gives a look at some exotic methods that are not widely available but are the tools of experimental physics. In the world of physics research, there are many types of radiation known or postulated to be produced, in places ranging from the university laboratory to the farthest reaches of the universe. It has become an immutable principle of this science that if a radiation type cannot be detected, then it simply does not exist.

DETECTION BY IONIZATION

Radiation was detected by ionization long before there was a theory to predict or explain it. Abraham Bennet (1749–99) was an English clergyman who happened to be interested in the new, fashionable scientific discoveries of his time. In 1787, he was particularly interested in electrical phenomena, and he concentrated on improving a method of measuring the presence of electricity called the *electroscope*.

Electroscopes had been around for nearly 200 years. William Gilbert, physician to Queen Elizabeth I, invented the first one, the versorium, consisting of a needle balanced on a pivot and first demonstrated in 1600. If charged with static electricity, the needle would then repel away from a similarly charged wand. This design was upgraded, or at least changed, in 1754, to a ball of pith suspended in midair by a silk thread. It could be attracted to or repelled from an electrically charged object, depending on the polarities of the charged pith ball and the charged object. Bennet's improvement on this dainty arrangement was dramatic. In his version of the electroscope, he hung two gold leaves from wires, facing each other, in a glass bottle. The conducting wires were connected, and they terminated in a metal ball on top of the bottle. Touch a static charge to the ball, as collected by stroking a cat or walking across a carpet, and the two gold leaves would spring apart from the mutual repulsion. Both were charged with the same polarity of electricity, and they repelled each other. The angle of separation between the two leaves indicated the strength of the charge. With this invention, the quantity of charge could be measured.

To improve the performance of the electroscope, subsequent versions were evacuated, to prevent leakage of the electricity off the gold

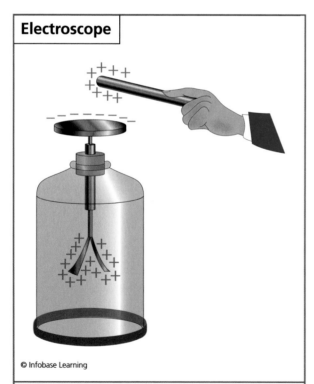

Electroscope

© Infobase Learning

In the classic electroscope, two hanging strips of metal foil repel each other and move apart as an electrical charge is applied.

leaves through air in the bottle. Still, with the route of electron leakage removed, the angular separation of the two featherlight leaves would decay as the charge diminished. The reason for this decay would remain a mystery until the end of the 19th century, when ionizing radiation was discovered. The electrons were being literally knocked off the gold leaves by radiation. Even without a lump of uranium ore near the electroscope, cosmic rays coming straight down out of the sky would eventually negate all the charge on the leaves. The invisible electromagnetic rays, past visible light on the wide spectrum of radiation, were able to ionize atoms, removing electrons from their natural positions, and this made the rays detectable in an indirect way. The rays were never visible, but the effects of the rays could be seen if one knew where to look.

The closing years of the 19th century were a flurry of activity in physics, with the rapidly developing discoveries of X-rays, then natural radiation, and then radium, the highly radioactive daughter product of uranium decay. The word *radiation* was coined by the discoverers of radium, Marie Curie and her husband, Pierre Curie. Pierre and his brother, Paul-Jacques Curie (1856–1941), had developed a piezoelectric version of the electroscope back in 1881. It was far superior to the antique models, and they upgraded the name to *electrometer*. It was capable of detecting small changes in electrical charge, and Marie used it extensively for measuring the presence, the intensity, and the decay rate of newly discovered radiations in the Curies' laboratory in Paris.

Research into the origin of radioactivity continued, particularly at the University of Manchester in England, where Ernest Rutherford (1871–1937) was overseeing a number of promising programs. Among his students and peers was Hans Wilhelm Geiger (1882–1945), a fresh physics doctorate from the University of Erlangen in Germany. With Rutherford in 1908, he developed a new type of electronic radiation detector. It would be named the Geiger counter.

It was a variation of the previously invented *ionization chamber*. The ionization chamber was simply a cylindrical metal can, with a wire running down the cylinder's main axis. The wire and its encircling can were electrically insulated from one another, and the can was filled with gas. Various gases were tried, including neon, argon, pure nitrogen, and plain air. An electric battery was connected across the two components, with the can charged positive and the center wire charged negative. Ionizing radiation entering the can would cause a line of broken atoms of gas to develop in its path. The electrons knocked off the atoms would then drift toward the wall of the can, and the now positively charged atoms would drift toward the center wire. This would cause a small electrical current to develop across these two electrodes.

The current in the ionization chamber caused by the passage of a ray or particle was quite small, on the order of 10^{-15} amperes, and it required a great deal of amplification to turn this phenomenon into a detectable signal. Still, the amplitude of the current, however small, was an indication of the amount of ionization occurring in the chamber on a moment-to-

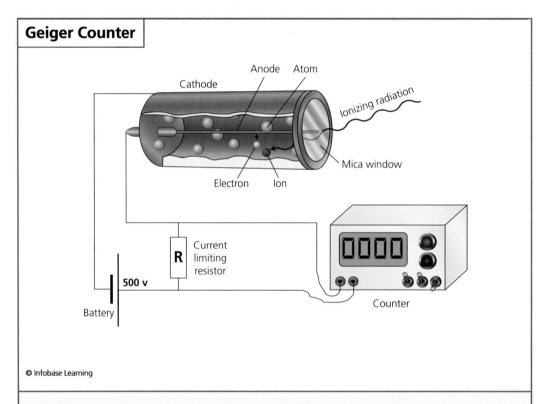

Geiger Counter

© Infobase Learning

The Geiger counter tube is a hollow, sealed metal cylinder filled with nitrogen gas. An electrical potential of 500 volts is established across the Geiger tube, from the inner anode wire to the outer cathode, by a battery or a direct-current power supply. When an ionizing radiation particle passes through the tube, it leaves a track of ionized gas particles, and the situation quickly becomes an avalanche, with all the gas ionizing at once. The resistor, R, in the circuit limits the current that can flow from the battery as the tube avalanches to a dead short. The event is quickly quenched by a small amount of halogen gas mixed with the nitrogen in the tube, and the gas returns to its normal, nonionized condition. Each ionizing event is counted by an attached device that detects the electrical current that flows momentarily through the tube.

moment basis, and therefore was a measure of the intensity of the radiation source being monitored.

Geiger had a hunch about the ionization phenomenon, and he turned up the voltage across the device, from 80 volts to 800 volts. The insulation was sufficient to allow this increase without arcing over in the chamber. With the voltage set at a critical level, a single ionizing event inside the can, instead of just forming a straight line of ions, would cause an avalanche. At the highly excited level of hundreds of volts, a recently ionized gas particle would not just drift toward the electrode, it would launch toward it, crashing into an adjacent molecule and causing further ionization. In a nanosecond, the entire contents of the chamber would be ionized, and a dead short would develop across the electrodes.

It was a brilliant move. This new mode of operation of the ionization chamber produced a strong, positive signal for each individual particle of radiation that traversed it. Instead of a barely detectable current caused by a stream of radiation, in this case very small radiation fields could be measured, counting the rays. For the experimental atmosphere of the early 20th century, it was a fortunate development. A purchased radiation source for experiments, such as commercially prepared radium-226, cost $28,000 per gram. Only tiny radiation sources were practical, and the vastly improved sensitivity of the Geiger counter was instantly appreciated.

There were improvements that could be made to the new Geiger counter. Unfortunately, once the chamber was shorted out by the ionization avalanche, it had to be unshorted, which meant that the voltage across the electrodes had to be cut and then reestablished for each ionizing event. Although this could be done electronically, there was a dead time in the detector, or a time in which no radiation counting could occur while the device reset itself. This would affect counting statistics, as radiation events occur with perfect randomness, meaning that two radionuclide disintegrations could occur only after a long delay, or at vanishing small intervals. The small intervals were hard to detect if the counter was dead while being deionized. The rate at which radiation could be detected was affected by the dead time, no matter how small it could be made.

Moving back to Germany, Geiger joined the Physikalisch-Technische Reichsanstalt in Berlin in 1912. He later in 1925 became a full professor at the University of Kiel. There he and his doctoral student Walther Müller (1905–79) found a simple and effective way to automatically quench the ionization in a Geiger counter in 1928. They found that a slight impurity of an organic compound, such as alcohol or ether, in the fill gas in the chamber would rapidly oppose the ionization and shut down the process. The resulting device could easily detect a traversing particle, indicate it as a click in earphones, and reset, waiting for another event, all within one nanosecond. It was possible to detect a billion events per second, if only there were a counter fast enough to do so. One could get a good impression of the radiation intensity by listening to the sound of particles hitting the Geiger tube in headphones or on an audio amplifier. At high count rates, the random clicking became a raucous buzz. At extremely high count rates, the Geiger counter would jam and go quiet. A rate meter was developed. Connected into the clicking signal, the rate meter would show the relative intensity of a radiation field as voltage on an analog gauge. They named the improved device the Geiger-Müller counter.

As this fine laboratory work proceeded, the industrial uses of radium-226 grew rapidly in the early 20th century, particularly in medical practice. Radium found uses for treating solid-tissue cancers as an inserted needle or applied in a bandage. Each radium needle was extremely small, usually less than a quarter-inch long, and they were amazingly easy to lose. Radium needles used in hospitals wound up on stair landings, in the gap between floorboards, behind toilets, in incinerators, in garbage pits, and, in one case, inside a pig. Radium needles were expensive and hard to get, with the entire supply in the United States peaking at about 11 ounces (300 gm). They were also the most dangerous radiation source that ever existed, and a lost radium needle was potentially harmful. A need developed for radiation-detection instruments for finding lost radium needles.

The first company to become fully organized for the manufacture of radiation instruments was the Victoreen Instrument Company, founded in Cleveland Heights, Ohio, in 1928 by John A. Victoreen (1902–86). Victoreen was an inventive radio amateur, but his life's ambition was to build and sell small, handheld instruments for monitoring X-ray machines. His first try was the Condenser R-Meter, built in 1927. It was an ionization chamber, mounted on the end of a metal wand, attached to a handheld aluminum box. To use it, the operator would turn a knob, which would charge the chamber with static electricity, and then read the radiation level by peering into a microscope atop the instrument. A sensitive electrometer under the microscope would indicate radiation hitting the ionization chamber.

Radium loss began to approach epidemic levels in the 1930s, and several detector devices were built with names such as "Radium Hound," "Radium Hen," and simply "Radium Finder." With costs increasing to $12,000 per radium needle in a depression economy, it became essential to locate lost units. By this time, the Victoreen improved R-Meter was selling well in the hospital equipment market. Always striving for increased sensitivity, hospitals began buying portable Geiger-Müller counters. In 1936, a Geiger-Müller counter was about the size of a suitcase and weighed more than 40 pounds (18 kg). The case had to contain a sufficient number of batteries to add up to 900 volts, as well as fragile vacuum tubes used as amplifier elements. It was a delicate piece of equipment, but fortunes in lost radium needles were recovered. About 40 percent of the lost needles, numbering in the hundreds, were never found.

With the beginning of World War II in Europe in 1939, the Victoreen Instrument Company was in a good position to supply radiation-detection equipment to the top secret U.S. government project to develop an atomic bomb, the Manhattan Project. Victoreen rightfully called itself the "First Nuclear Company." By late 1942, there was a big need for portable instruments for field use. The Manhattan Project needed Geiger-Müller counters to monitor radioactive sources, contaminations on floors or in outdoor experiments, and in hundreds of experiments involving small radiation emissions. A handheld Geiger-Müller counter could easily read radiation as slight as 0.01 milliroentgens per hour, and it was sensitive enough to detect dangerous radiation at a distance.

Also needed were *survey meters,* portable, handheld ionization chambers with analog meter readouts in roentgens per hour. A range switch could take the sensitivity down to hundreds of roentgens per hour, or a seriously hazardous dose rate, certain to cause physical problems. Ironically, with the new development of powerful fission reactors and nuclear explosives, a highly sensitive Geiger-Müller counter could be quickly overwhelmed by the radiation intensity now available. For the sake of safety, technicians were advised to start a radiation sweep with an ionization chamber, looking for a dangerous condition, starting with the highest detection range, in hundreds of roentgens per hour, and then switch down to the lower ranges. There was danger of a hazardous radiation environment overload if detecting in a low range.

Ingenious new detectors were designed at Los Alamos Laboratories in New Mexico, the Metallurgy Laboratory in Chicago, and the Clinton Works in Oak Ridge, Tennessee, and many designs were passed along to instrument manufacturers such as Victoreen for limited mass production. One of the most successful designs from the bomb project was the "Cutie Pie" handheld ionization chamber. The user would hold it by a pistol grip, point the cylindrical ionization chamber on the front, and look down at a meter attached to the top. It quickly found use surveying reactor faces,

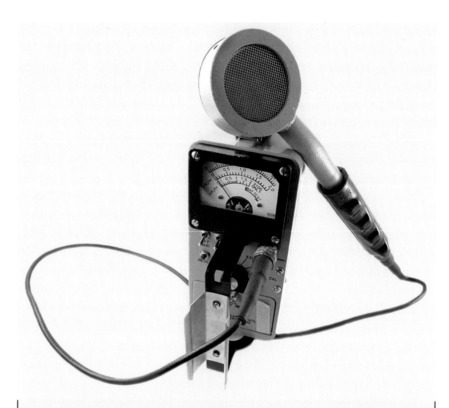

A Geiger-Müller meter, with a pancake probe attached *(Ilya Rabkin, 2010, used under license from Shutterstock, Inc.)*

"glory holes," and irradiated items for dangerous radiation. The Cutie Pie, named for the response of a scientist when he first observed it, is still in use in every nuclear power and research facility in the world.

Gamma rays were relatively easy to detect with an ionizing radiation detector of any type, as they were sufficiently powerful to penetrate the detector's outer electrode. Beta rays were more difficult, as they lack penetrating power, and thin-walled detectors were developed specifically for beta rays. Alpha particles have very slight penetrating power, and special equipment had to be designed to detect these unusually dangerous rays. In the earliest experiments, it was found that an alpha-emitting source placed inside the detector would defeat the problem of penetration. Dynamic gas detectors were developed to detect alpha sources carried in air, as air was used as the fill-gas and was blown through the detector cylinder. At Argonne Laboratories in Illinois in 1947, the Pee Wee Mark

1 Model 41 alpha portable survey meter was developed, using a very thin sheet of aluminized Mylar film as one wall of the ionization chamber. The Mylar was thin enough to allow alpha particles to enter the chamber, and the aluminum covering, a scant few molecules thick, conducted electricity. Held against an alpha particle source, the meter would evaluate the intensity of the alpha emissions.

The Pee Wee was a new class of ionization detector, the proportional counter, and it would transform the radiation-counting capabilities for all types of ionizing radiation. A Geiger-Müller counter responds strongly to a single radiation particle, but its response is the same for all energies and types of radiation. An ionization chamber also responds in the same way to all types of radiation, simply giving a measurable current that corresponds with the rate at which it is being hit with radiation particles. A proportional counter reacts to individual particles, but each reaction is proportional to the energy of the incoming particle. This gives another dimension to the radiation counting, and the type and danger potential of particles can be derived from the energy that each deposits in the detector tube.

A proportional detector tube is similar to a Geiger or ionization chamber tube. It is filled with a mixture of 90 percent argon and 10 percent methane gas. Between the relatively low voltage applied to an ionization chamber, 50 to 250 volts, and the Geiger voltage, 800 to 900 volts, is the proportional region. In this range of voltage, an ionizing radiation event in the fill gas causes a cascade of further ions to be produced in what is known as the Townsend avalanche. Unlike the Geiger avalanche, the Townsend effect creates a number of additional charged particles that is proportional to the energy of the initiating event, instead of shorting out the entire tube. The signal from a Townsend event can be amplified and used for subsequent analysis.

After World War II, the need for portable radiation-measuring equipment of all kinds expanded rapidly. In 1946, Victoreen's government sales were $800,000. In 1949, the sales were $4,500,000. In 1946, Victoreen produced the Model 263 portable Geiger-Müller counter, and it was used in Operation Crossroads, the atmospheric atomic weapons tests at Bikini Atoll in the Marshall Islands. Users found it fine but fragile, describing it as "adequate on the desk of a meticulous old lady in an air conditioned laboratory, but not useful in boats." On military insistence, it was made more rugged, waterproof, and heavier.

With the development of the nuclear-powered submarine in the early 1950s, the United States Navy needed an entire range of radiation-detection equipment, from fixed dynamic gas detectors placed in locations over the entire boat to handheld detectors for designated crewmen to monitor the slightest leakage of radiation from the submarine's power reactor. Radiation safety was a primary concern of the new nuclear navy, and the only way to ensure it was to constantly monitor every potential radiation hazard. The army and air force developed radiation-detection equipment as well. High-altitude aircraft carried radiation counters aloft to analyze atmospheric effects of atomic bomb tests over a 20-year period, starting in the late 1940s. Helicopters carried radiation counters to probe for ground contamination. Tanks and shielded tracked vehicles were specially equipped for close detection of atomic bomb debris. For these military applications, gas-filled detectors of all types were used.

Gas-filled detectors using the ionization effect of radiation remain in daily use around the world, wherever there is nuclear power, fuel processing, waste storage, weapons, or research. Within the atomic bomb development program, starting in 1943, a new branch of science was created. It was named health physics, and its goal was to prevent anyone working in the nuclear industry or in nuclear research from being exposed to hazardous quantities of radiation by setting scientifically determined maximum levels of radiation dosage in all situations and monitoring these dosages using a wide range of measurement equipment. Radiation detection by ionization in gas immediately became an important tool for this discipline, and it remains so.

DETECTION BY SCINTILLATION

Radiation detection by measuring the ionization of gas is a well-established method, but it may not be the most efficient process. Gas is, by its nature, a thinly dispersed collection of molecules, consisting mainly of empty space. Gas is not dense, as is solid material, and the probability of a radiation particle actually making contact with a molecule as it travels through a volume of gas is not very good. Every process in nuclear physics seems based in probability, and the interaction of radiation with matter is no exception. The probability of a radiation-matter interaction, and therefore a detection, is simply higher in a liquid or a solid than in a gas. There is another way that radiation can affect matter called *fluorescence*, or scintillation, and it happens in certain liquids and solids.

In this process, the incoming particle or ray is moving fast and has a high vibratory frequency that makes it impossible to see. Human eyes can only make out electromagnetic radiation in the range of red to violet light, which is a tiny portion of the spectrum, and it is below the threshold of ionization. In fluorescence, the ray actually experiences a down conversion upon collision with an atom, in which its extremely high frequency is stepped down into the sub-spectrum of visible light. Upon interaction with certain chemical substances, called phosphors, the invisible radiation becomes visible, one ray at a time.

In 1903, scintillation effects were well known, and some of the first discoveries of invisible radiation in the last decade of the 19th century had depended on laboratory experiments using phosphors, such as zinc sulfide. Zinc sulfide glows a yellow-green color when bombarded with ionizing radiation and is visible to dark-adapted eyes. William Crookes (1832–1919), a British experimental physicist, created a new laboratory instrument using this principle. One day, at his private laboratory at 7 Kensington Gardens, London, he was observing the uniform glow on a phosphorescent screen imparted by a sample of powdered radium salt. He managed to spill some of the salt on the screen. He was able to pour most of it back in the bottle, but radium salt was so expensive he hoped to recover every tiny crystal of it. After brushing all that he could see into the bottle, he decided to use a microscope and look for any remainder. Looking at the screen in the dark under magnification, he was amazed. He was no longer seeing the dim glow that scientists had seen for the past eight years. He saw individual flashes of light. As single radium-226 atoms spontaneously disintegrated, he could see the flash as an alpha particle hit the zinc sulfide.

Crookes immediately saw the value of this observation. He closed one end of a brass tube, painted the inside with zinc sulfide, and inserted a microscope eyepiece in the other end. With a tiny speck of radium on the tip of a needle in the tube, he could watch the rate of radiation release with great accuracy, counting the number of disintegrations happening in a fixed length of time. He named it spinthariscope, for the Greek word meaning "spark." As a scientific instrument, it gained immediate use, for its ability to quantize radiation release down to individual disintegration events. Ernest Rutherford, the great experimentalist from New Zealand, used spinthariscopes at the University of Manchester in radiation-counting experiments to discover the atomic nucleus in 1909.

The spinthariscope was renamed the scintillation counter and became indispensable in the new science of nuclear physics. The phosphor screen was kept as thin as possible, so that a scintillation counter would detect only alpha particles. High-energy gamma rays were likely to skip through the screen without significant interaction, leaving a much weaker flash of light than a robust alpha that was completely absorbed by a thin phosphor. Research technicians would sit for hours, squinting through the microscope eyepiece and counting flashes of light against a timer set for one hour. It was exceedingly dull work, and the counting personnel would quickly become bored with the assignment. The human eye is a weak and inconsistent sensor. The mind of the counter could start making up light flashes if the count rate were too low. As accuracy of measurements gained importance, scientists tried to develop more reliable means of counting alpha particles. Discoveries in nuclear physics made using scintillation counters suspect, and by the mid-1920s photographic film had replaced most scintillation counter applications. The film left an indisputable, permanent record of radiation, not subject to an individual's imagination. By 1930, the use of a scintillation counter for radiation measurement was rare.

The basic principle of using fluorescence for radiation detection was sound, but there had to be a way to remove the human element from the measurement for it to have further scientific use. It was suggested as early as 1933 by Berta Karlik (1904–90), a physicist in Vienna, Austria, to electronically measure the light flashes, using a photoelectric vacuum tube, but the photo sensors at the time were insufficiently sensitive to detect such dim scintillations. In 1934, a breakthrough in electronic light-detection techniques came to the attention of the Radio Corporation of America (RCA).

RCA had sent the head of its Camden, New Jersey, research laboratory, Vladimir Zworykin (1889–1982), to the Soviet Union on a sales trip. Zworykin was a Russian-American engineer and inventor, credited with pioneering aspects of television, and RCA was trying to sell radios in the large, untapped market of Soviet Russia. Zworykin impressed the Soviets with the latest American inventions in electronics, and the Soviets brought out their latest invention. Leonid Aleksandrovitch Kubetsky (1906–59), a physicist working in Leningrad, had just perfected a multistage photomultiplier tube, using a silver-oxygen-cesium photocathode. It was impressive, giving a thousandfold amplification of light hitting the end of the tube. Kubetsky's tube was capable of detecting a single photon of light and converting it to a pulse of electrical current. Zworykin was quietly impressed and made a diagram of what he had seen as soon as his party was clear of the Soviet Union on hotel notepaper in Berlin, Germany, on the way home.

Back in New Jersey, RCA immediately began long-term development of this device, which they named the photomultiplier tube. With improvements, the sensitivity of a photomultiplier increased by 100 million, and this new technology found immediate use in astronomy, to measure and record the action of pulsating stars. During World War II, all electronic research was turned to the war effort, and RCA's photomultiplier found an unexpected application. It was used as a noise generator in radar-jamming devices.

After the war, in 1946, Marietta Blau (1894–1970), an Austrian-American physicist working in industry, suggested a reapplication of the long-lost art of scintillation counting. She modified Berta Karlik's suggestion, which she remembered from 1933. The RCA photomultiplier tube could replace the human eye as a detector, amplifying the extremely small light signal into an electronic signal of such magnitude it could be used to drive a meter or an electronic counter. Robert Hofstadter (1915–90), a professor of physics at Stanford University in California, is credited with the reinvention of the scintillation counter in 1947.

Hofstadter's improvements on the scintillation counter concept were profound and led to a type of radiation instrument that remains in common use in nuclear research and nuclear power applications all over the world. Instead of interacting with alpha particles on a thin coating of zinc sulfide, Hofstadter used a polished, cylindrical block of sodium iodide, laced with a small

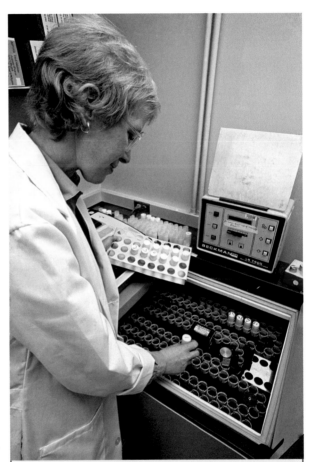

The use of a scintillation counter for measuring radiation at the Stanford Medical Center in California. In this special case, the scintillator is a fluid, contained in small, disposable plastic cylinders. The radioactive sample to be counted is dissolved in the scintillator, and individual cylinders are moved on a track to face a photomultiplier tube for evaluation of the radiation level. *(Peter Arnold, Inc./Alamy)*

amount of thalium used as an activator. With the sodium iodide optically coupled to the sensitive end of a photomultiplier tube, individual, high-energy gamma rays could be counted electronically, with an efficiency 200 times that of the best Geiger-Müller counters. In traversing a solid block of transparent sodium iodide, a gamma ray was very likely to run into atoms and cause the down conversion into a spark of light, particularly if it managed to hit the crystal in the center. Gamma rays shooting through a corner or an edge were less likely to scintillate, so the efficiency or sensitivity of the device depended on the size of the material. Depending on the amount of radiation sensitivity that was desired, the block of sodium

UNUSUAL DETECTION METHODS

There are hundreds of different types of exotic radiation detectors, from barium platinocyanide scintillation counters to direct ion storage counters, usually invented as the need arose for more advanced nuclear physics investigations. Ever larger high-energy nuclear experiments probing the fundamental construction of matter are being built, such as the Large Hadron Collider (LHC) at the Organisation européenne pour la recherche nucléaire (CERN), located near Geneva, where it spans the border between France and Switzerland about 100 m underground. A proton is an example of a hadron, and the LHC crashes them into each other at extremely high speed. This powerful nuclear particle accelerator, operating at a proton energy of seven tera-electron-volts, is equipped with six advanced particle detectors.

An example is ATLAS (A Toroidal LHC ApparatuS), which is a general-purpose detector 144 feet (44 m) long, weighing 7,716 tons (7,000 mt). It was invented by a team of roughly 2,000 scientists and engineers at 165 institutions in 35 countries. In addition to detecting fundamental particles erupting from a hadron collision, it is a muon spectrometer and a hadronic calorimeter. Raw data from the detector is about 25 megabytes per event, with 40 million events per second expected during a collision. It is set up to detect microscopic black holes, Hawking radiation, and the Higgs boson. This seems a far cry from William Crookes developing the spinthariscope in an afternoon, alone in his room overlooking Kensington Gardens.

ATLAS is designed to track the path of particles traveling through it in three dimensions, but this is not as advanced as it may seem. The first such detector was the Wilson cloud chamber, developed by Charles Thomas Rees Wilson (1869–1939), a Scottish physicist, beginning in 1894. Wilson was studying the optical effects of clouds at the summit of Ben Nevis, the highest peak in the British iodide could be any size, from a stack of 10 quarters to an entire room. The new scintillation counter, or scintillator, found applications in cosmic ray research, fundamental nuclear physics, medical imaging, homeland security, and uranium prospecting. Hofstadter won the Nobel Prize in physics for his work using scintillators in 1961.

Many variations of the scintillator have been developed. Sodium iodide with thalium is used to detect gamma rays. Cesium iodide is used to detect alpha particles and protons. Clear plastic, usually containing anthracine, is used for large scintillation blocks, and even organic liquids can be used for scintillators. The radiation source to be counted

Isles, in Lochaber, Scotland. He decided to build an expansion chamber that would produce model, artificial clouds on a desktop and was successful. He found that the microscopic water droplets making up a cloud have to form on something, and, to his surprise, clouds will form on the ion trails left by radiation traveling through the air. His chamber was cylindrical, with a round glass window on top so he could observe the clouds. With the air in the chamber saturated with water vapor, Wilson would suddenly decrease the pressure with a quick pull on a pump handle. Cosmic rays or a small radiation source, such as a speck of radium-226, would cause vapor trails, cutting through the three-dimensional space of the chamber. He perfected the device in 1911 and won the Nobel Prize in physics in 1927 for having thought of it.

Alpha particles in the cloud chamber make thick clouds, and beta rays make thin clouds. Place a magnet under the chamber, and electrons from β- decay can be easily distinguished from positrons produced by β+ decay. The electrically charged particles are influenced by the magnetic field. Electrons spin in a left-hand spiral, and positrons spin in a right-hand spiral.

Another Nobel Prize in physics was won in 1960 by Donald A. Glaser (1926–), professor of physics at the University of California, Berkeley. His winning idea was to use liquid hydrogen instead of water vapor in a device similar to the cloud chamber. If the hydrogen were at just the right temperature and pressure, then it was on the verge of boiling. Shooting a particle of radiation through the volume of hydrogen caused it to boil to hydrogen vapor in the track. As the particle burrowed through, it left a trail of tiny bubbles. A magnet coil around the chamber causes the bubble path of charged particles to curl, and the speed of a particle can be derived from measuring the curvature. The bubble chamber is still in use for specialized investigations and was used recently to detect a rare subnuclear particle called the weakly interacting massive particle (WIMP).

can be dissolved in the scintillator fluid, ensuring the collection of every ray emitted. Low-energy beta rays from tritium and carbon-14 are so counted.

As nuclear security concerns have increased in the 21st century, the use of large-scale, industrial scintillation counting has become commonplace. Scintillators are now used in ports, weigh bridges, and scrap metal yards to check bulk shipments for inappropriate radioactive materials and atomic bomb components, still using the photomultiplier tube design borrowed from the Soviets by Zworykin in 1934.

SEMICONDUCTOR DETECTORS

Solid mass radiation detectors, such as the scintillation counter, are very useful for gamma-ray and X-ray spectrometry. A disadvantageous factor is that the molecules of such a solid are in constant, thermal motion, always wiggling. As a radiation particle hits a molecule, its energy is perceived not as its incoming velocity, but as the sum of the particle velocity plus the velocity of the molecule it hits. If the constantly moving molecule happens to be moving toward the incoming particle, then the perceived energy is artificially high, and if the molecule is moving away from the particle, then the energy is seen as lowered. This Doppler effect blurs the measurement of particle energy, and rays of similar energy tend to melt together on a diagnostic graph.

If only the solid molecules could be made to stand still, the measurements would snap into focus and a spectrum representation would be sharp and clear. The temperature of a material determines the speed of the molecular jiggling, and if the scintillating block could be cooled down far enough the motion would effectively stop. Unfortunately, a cold scintillator will not scintillate. The ability to make sparks of light due to radiation disappears at low temperatures.

Semiconducting solids, such as silicon and high-purity germanium, can detect ionizing events in their crystalline matrices of molecules using the ionization principle, with electrons broken free by the passing of energetic particles and drifting toward the positive electrical terminal. This is a solid-state ionizing effect and not a fluorescence as used in scintillation counting. It works best at the temperature of liquid nitrogen, which is $-320.15°$ Fahrenheit ($77.36°$ K), and at this temperature the molecular motion is slight. Gamma-ray and X-ray spectrometry is sharp and precise under this condition. Handheld spectrometers for use in the field are still

This microsensor uses neutrons to detect organic contraband. *(AP Images)*

scintillator based, but for laboratory work where liquid nitrogen can be used, nothing outperforms the fine resolution and clarity of an intrinsic germanium detector.

It is also possible to use solid blocks to detect radiation using no electricity at all. Solid-state nuclear track detectors, or etched track detectors, use the fact that radiation particles traversing through a solid will weaken the internal structure along the track. Subject the solid to an etching solution, acid or base, and the damaged molecules will etch out faster than will the untouched molecules, and visible pathways of flying particles are left in the material.

The track detector is preferably transparent, so that the etched tracks may be seen. Examples of etched track materials are photographic emulsions, crystals, glass, or plastic blocks. An often used plastic is polyallyl diglycol carbonate, also known as TASTRAK. Etched track detection is used to study cosmic rays, long-lived radioactive elements, and radon gas in basements. It has been used to date the age of geological samples. The space helmets of Gemini orbital astronauts in the 1960s were etched and examined under microscopes to determine the cosmic ray exposure of men in extended spaceflight.

NEUTRON DETECTION

Neutron detection is important, particularly for nuclear power operation, where the concentration of neutrons is a fundamental measure of reactor power. It is also used for radiation protection monitoring in nuclear power situations, for nuclear and cosmic ray research, and for bomb detection in airports. There are many ways to detect neutrons, most using the same methods as are used for ionizing radiation detection in an indirect way.

The detectors used for power monitoring in the first successful nuclear reactor experiment at the University of Chicago in 1942, the CP-1 experiment, remain in constant use. These were *BF3 detectors*. The BF3 is a Geiger-Müller or proportional ionization chamber using boron tri-fluoride as a fill gas. The boron is a special nuclide, boron-10, which is an isotope separated from naturally occurring boron. Natural boron is only 20 percent boron-10. For neutrons traveling at thermal speed, the neutron absorption cross section of boron-10 is enormous, at 3,840 barns. Upon capture of a thermal neutron, boron-10 immediately endures alpha decay into lithium-7. The energetic alpha particle, generated inside the Geiger-Müller tube, causes a distinct electrical signal.

This is a fine detector for thermal neutrons, but if high-speed neutrons are to be detected, then they must be slowed, usually by a covering blanket of paraffin or plastic. These materials contain hydrogen, which will slow the incoming neutrons by collisions with the still protons of the hydrogen nuclei.

For special use in detecting fast neutrons, proton recoil is exploited. A Geiger-Müller tube filled with hydrogen will detect fast neutrons as they collide with the hydrogen nuclei, or protons, and send them off at high speed, ionizing the gas and causing countable electrical events. The proton-recoil principle is also used in neutron scintillation counters for fast neutrons. An example of a scintillator material used for neutron detection is zinc sulfide, the phosphor once used in spinthariscopes, dissolved in paraffin. The paraffin supplies protons for recoil, and the zinc sulfide scintillates when hit with a launched proton.

The simplest way to detect neutrons has been used since the discovery of nuclear energy, and it involves simply metal foils. Called activation detectors, small squares of foil are placed in strategic locations in a nuclear reactor or experiment to map the neutron flux profile. Depending on the sensitivity desired, the foil can be any of a number of possible metals. Indium, gold, manganese, vanadium, and cobalt are used. After being exposed to the neutron activity in a reactor for a specific length of time, the foils are removed and the induced radioactivity from neutron activation is counted under laboratory conditions. This measurement is accurate and indisputable.

The impressive range of devices used to detect neutrons, a neutral particle that can wander through matter looking like just another molecule, seems to indicate that if a subatomic or subnuclear particle exists, then human ingenuity will find a way to detect and measure it.

DOSIMETRY

Dosimetry is the industrial art of measuring the radiation dose accumulated by individuals in the fields of nuclear power, medical care, scientific research, and wherever radioactive materials are used. An instrument used to record radiation doses is called a dosimeter.

The first dosimeters, used as early as 1922 for X-ray technicians, were film badges. A film badge is simply a small rectangle of unexposed photographic film, enclosed in a light-tight wrapping or container, clipped to the user's shirt pocket. Ionizing radiation hitting the dosimeter will expose or fog the film. At the end of a workweek or after an incident where radiation dosage is suspected, the film is chemically processed and evaluated by a light densitometer. The density of the blackening of the developed film is proportional to the received radiation dose, and it is a permanent record that can be filed away under the individual's name. The film badge is mechanically simple and requires no batteries or adjustments. Although no longer in wide use in the United States, film badges are still used throughout the world in nuclear industries.

Film badges were fine for making a permanent record of one's radiation exposure, but the user could not measure his or her dosage as it occurred. The only way to read a film badge dose was after the fact, after the possibly damaging radiation exposure had occurred. To remedy this situation, the direct-reading pocket dosimeter was developed. It is an aluminum tube, to be carried clipped in a shirt pocket, about the size and shape of a fountain pen. The amount of radiation that has impinged on the dosimeter can be read at any time, simply by looking through the end of the tube while pointing it toward a light source. Visible through the eyepiece at the top of the unit is a lighted scale, marked off in roentgens or milliroentgens, with a fine, black line through the measured quantity. Each Monday morning the dosimeter is charged, with its pointer run back to zero, by plugging it momentarily into a charging station.

The mechanism within the aluminum tube is simply a radiometer, working on the same principle as the ancient electroscope. A quartz fiber is used instead of the traditional gold leaf, and when pointed at light it throws its thin shadow across a frosted glass with a scale etched on it. The charger puts about 100 volts on the fiber, running it back to the zero position on the scale and giving it an electrical charge that can be influenced by incoming radiation. The pocket dosimeters give nuclear workers a reassurance that they are not in immediate danger by holding it up to the light occasionally and reading the lack of dose on the indicator screen. A rap on a concrete floor, however, can skew the reading and cause undue concern at the huge, but false, indicated radiation dose.

The film badge and the pocket dosimeter, long mainstays of the nuclear field, have been widely replaced by an even simpler mechanism, the thermoluminescent dosimeter (TLD). A TLD is simply a crystal of either calcium fluoride or lithium fluoride, small enough to fit on the end of a pencil eraser. It is held in a plastic badge, which is clipped to the clothing and worn in radiation environments. Each badge is serial numbered and linked to the wearer's name in the computerized record-keeping system. At the end of the week, or more often in cases of unusual radiation work, the crystal is removed from the holder and placed in a

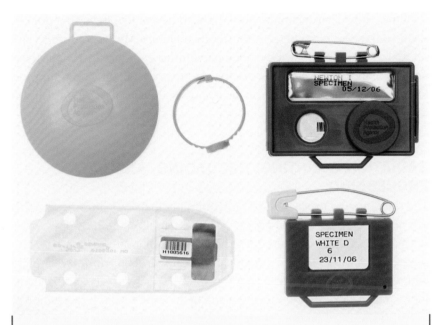

The types of personal radiation dosimeters seen here are: radon radiation detector (yellow, upper left); thermoluminescent dosimeter (TLD, purple, upper right); and a neutron dosimeter (green, lower right). The ring-shaped and finger glove dosimeters (upper center and lower left) are variations on the TLD design. *(Health Protection Agency/Photo Researchers, Inc.)*

TLD reader, consisting of a small electric heater in a light-tight container with a photomultiplier tube to evaluate light coming from the crystal. The electronic signal from the photo-detector is automatically loaded into the TLD wearer's permanent digital record.

The TLD effect is a variation of the scintillation counter. In a scintillation counter, a block of sodium iodide will scintillate under radiation unless it is cold. The sliver of lithium fluoride in a TLD will absorb the energy from radiation, but it will not light up until it is warmed with a heater. All the radiation exposure is held in the crystal until heated, when a light-sensitive photomultiplier is used to evaluate the number of light photons given off, which is proportional to the radiation absorbed. These devices are used by personnel in every nuclear power plant, fuel-processing facility, and nuclear research laboratory in the United States. They are attached to barrels of waste, equipment, light fixtures, walls, rocks, and trees, to monitor radiation exposure in every application of radiation. They are highly reliable.

Other dosimeter technologies have been developed, including pocket-sized ionization chambers, subminiature Geiger-Müller counters with set alarm thresholds, and even Geiger-Müller counters in watches, but nothing matches the stark simplicity of a tiny crystal of lithium fluoride. For the continued safety of nuclear workers, the TLD will gain usage worldwide.

RADIATION COUNTING, RECORDING, AND DATA PROCESSING

The fundamental processing of radiation data involves counting ionizing events, usually over a predetermined time interval. In the early days, when scientists and their students spent hours squinting into microscope eye-pieces, one would hold a mechanical tabulator in one hand, punching it with the thumb every time a spark of scintillation was seen, and waiting for an alarm clock to announce the end of an hour's time. It was tedious.

With the introduction of electronic instruments, such as the Geiger counter, there was an immediate need for an electrical counter to total up the number of radiation events, again using an alarm clock. Electro-mechanical decade counters were made in the late 1920s using magnetic telephone stepping relays. These special devices were used in the new Bell System automatic telephone exchanges, but they also found use in nuclear research. Each pulse directed into the low-digit relay would bounce the relay arm forward one notch. After it had gone 10 notches, the relay arm would run back to the zero position and send subsequent pulses to the next relay in line. Using a rack full of these things, a researcher could count pulses without the error-prone involvement of a graduate student with a drifting mind, and an alarm clock would automatically turn off the assembly after one hour of counting.

This was a vast improvement over the previous system, but it was too slow to handle the ever-increasing rates of radiation counting experienced in the 1930s. By the time of the World War II crash program for atomic bomb development, much faster counting means were developed. Fully electronic counters, or scalers, were developed for precise counting of radiation, at speeds as high as a billion counts per second. Neon lamps, having unique, digital electronic properties, replaced the 10 positions of the telephone relay arm, lighting up when pulsed and passing the circuit to the next lamp in line. Lines of 10 lamps would cascade into adjacent lines of 10 lamps, forming decades of counting capacity. The response of a neon counter was quick, at about one nanosecond. At the end of a counting session, timed by a mechanical clock, a research technician could then read off the number of radiation events on a panel of neon lamps.

Developments in nuclear physics in the decades after the war were in direct parallel with the evolution of the digital computer, and the two technologies fit perfectly together. A great deal of nuclear work was needed to find and characterize hundreds of newly discovered radioactive nuclides, each of which had multiple radiations of different energies, types, and half-lives. To measure the energy of a particle or ray and to distinguish it from others in a cacophony of radiation, it was necessary to determine not only the rate of decay by counting but the energy of each radiation event. For this purpose, the single-channel analyzer was developed. Pulses from an energy-sensitive detector, such as a scintillation counter, were fed into an analyzer gate. This gate had two settings, low threshold voltage and high threshold voltage, defining a channel. Only a pulse from the detector with a height, or voltage, within the upper and lower limits was permitted to proceed to the counter circuit. Therefore, only radiation events of a certain predetermined energy were counted, with everything else disallowed. William "Willy" Higinbotham (1910–94), an American physicist, built one of the first single-channel analyzers, the Model 200 Pulse counter, in 1947, just in time to be used with the newly developed scintillation counter. Higinbotham was head of the instrumentation division at the Brookhaven National Laboratory on Long Island, New York, and he is best known for the invention of the digitized tennis game called Pong in 1958.

This ability to count only particles at a specific energy was a step forward, and it had been accomplished with vacuum-tube technology. The next step would be to expand the technology to two, three, four, or more channels. This would allow counting of more than one energetic event at one time, and it would simplify searches for radiations of unknown energies. Using just one channel, it was necessary to start with a wide acceptance window and then scan and narrow down the thresholds with subsequent counting sessions. It could take days to pin down the characteristics of a single nuclide emission. Using the rapidly developing digital technology used in electronic computing machinery, it was possible to build a 99-channel pulse analyzer by 1950, using analog-to-digital conversion techniques. By 1955, magnetic core storage of channelized counts was possible, and by 1956 a 256-channel analyzer was built at the Argonne National Laboratory in Illinois.

With 256 pulse-height or radiation event energy channels, it was possible to count an entire spectrum of energetic events in a single session, building a graph showing the number of events counted versus energy of an event. This was the spectrum of gamma-ray decay, displaying the energies and the relative quantities of multiple gamma-ray sources. Individual nuclides were easy to pick out of a sample containing multiple sources of radiation, and a nuclide could be identified by its characteristic gamma-ray energy. Data were transferred to digital computers, usually by punched data cards, to fit exponential curves to single-channel data decreasing with time, calculating the half-life of a specific nuclide. In 1956, a 256-channel pulse-height analyzer took the floor space in a large room having aggressive air-conditioning, to prevent the equipment from melting. A great deal of heat was generated by tens of thousands of watts of electricity required by the vacuum-tube electronics. The graph of number-of-counts versus pulse energy was plotted on an oscilloscope, similar to a television screen. The digital computer used to fit exponential curves, which was also vacuum-tube based, required an even larger space. The processing of nuclear decay data was rapidly advancing, as electronic computing gained speed and data storage ability.

In the 1960s, transistors and then integrated circuits replaced vacuum tubes in digital circuits, and both computing and multichannel analysis improved rapidly. The size of a 1,024-channel analyzer was reduced to an instrument with the bulk of a suitcase, and the counting equipment necessary to identify unknown radioactive sources became semi-portable. Industrial and scientific minicomputers, reducing the size of a general-purpose computer to one or two cabinets of equipment, brought together the counting and the computing functions into a single unit. Given the power of this combination, it became possible to store observed and identified gamma-ray energies in a computer-driven analyzer system.

Today, an entire multichannel pulse-height analyzer system is in a device that fits in the palm of the hand, no larger than a cell phone. Equipment development has come a long way since the 1920s, but the goal remains constant: Find radiation where it occurs, determine its intensity, and identify the source by its energy. As nuclear power, nuclear medicine, and nuclear industry become a larger part of human activity, these measurements increase in importance. In the next chapter, measures to avoid radiation and its effects are detailed.

6 Radiation Avoidance and Protection

Following closely behind the discoveries of radiation types and ways to detect them came an awareness that excessive exposure to radiation could modify equipment, change the normal path of chemical reactions, and, to a larger extent, harm living things. Anything with a spark of life, from single-cell organisms to laboratory workers, could be adversely affected by large doses of radiation. From these observations, the next step in an understanding of radiation and its effects was to formulate ways to avoid it and to implement plans based on this new knowledge. The result is a technical world that is capable of working with enormous measures of radiation of all types, from ultraviolet light to high-speed neutrons, without death, injury, or a measurable radiation effect on the workers, the surrounding area, or anyone in range of such a facility. The radiation industry, after a century of study and development, has become the safest workplace on Earth. Statistically, there is a greater chance of death or injury to stockbrokers and real estate agents than to nuclear power workers.

The entire canon of radiation exposure avoidance can be summed up in three words and enclosed in a single sentence. In this section, these three important concepts will be revealed and expanded, with history, explanatory text, and examples, to fill an entire chapter. Specific ways to monitor and find radiation sources and to avoid internal exposure are part of the expansion of the basic rules in this chapter. The sidebar on pages 109–110 gives one example of a lack of radiation protection and its effects in a famous incident following World War II.

TIME, DISTANCE, AND SHIELDING

Radiation sources have been as small as a microscopic bit of radium-266 held on the point of a pin to as large as a hydrogen bomb explosion enveloping a South Pacific island. Regardless of its size, a person's exposure to radiation depends upon the time spent in its presence, the distance between the person and the source, and any solid material, or shielding, that stands between the source and the person. Time, distance, and shielding are the three methods of protection from radiation dosage. To reduce the integral dose from a source of radiation, spend little time with it, stay a distance from it, get behind something, or do all three. A reduction in dose is an assured outcome even if only one of the three parameters is exercised sufficiently.

Suppose you are in the presence of a one microgram sample of cobalt-60. A microgram is an extremely small portion that requires a microscope to see. It is, essentially, invisible. Cobalt-60 emits an energetic, penetrating gamma ray with an energy of 1.33 MeV. This nuclide decays with a half-life of 5.28 years, or 1.67×10^7 seconds. Its decay constant, λ, is 4.14×10^{-8} disintegrations per second per atom. One microgram of cobalt-60, or about 1×10^{13} cobalt-60 atoms, releases gamma rays, or decays, at a rate of about 10 millicuries (3.7×10^7 Bq).

The radiation exposure received at a distance of an inch (2.5 cm) from this source over the course of two hours is 29 rads (0.29 Gy). This is a significant dose and will likely cause radiation sickness symptoms. Increase the exposure time to one day, and the dose at an inch is 350 rads (3.5 Gy). The exposure is directly proportional to time spent. Reduce the time to 0.2 hours, or 12 minutes, and the dose reduces to 2.9 rads (0.029 Gy). Dose accumulates with multiple exposures. If the one milligram cobalt-60 source has fallen to the floor and a radiation worker walks over it 100 times, each time spending one second within an inch (2.5 cm) of it, through the sole of a shoe, the bottom of the worker's foot receives an accumulated 4.0 rad (0.04 Gy) dose, with each passing dose being only .04 rads (0.4 mGy). The sole of the shoe provides inconsequential shielding for the 1.33 MeV gamma rays.

If this same 10 millicurie cobalt-60 source were three feet (one m) away, the dose for two hours would be reduced dramatically, to

0.026 rads (0.26 mGy) from 29 rads (0.29 Gy) for the same time spent an inch (2.5 cm) from the source. Move to 33 feet (10 m) from the source, and the dose for two hours drops to 0.26 millirads (2.6 µGy). The dose

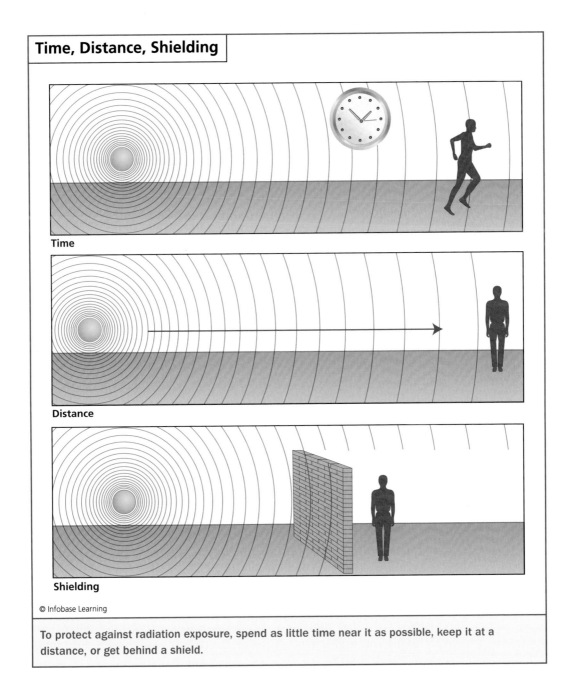

Time, Distance, Shielding

Time

Distance

Shielding

© Infobase Learning

To protect against radiation exposure, spend as little time near it as possible, keep it at a distance, or get behind a shield.

rate is inversely proportional to the square of the distance from the radiation source.

There is another strategy for reducing radiation exposure. Think of the visible light radiation from the Sun. An excellent shield from sunlight is an overhead roof. For mobile, outside activities, a large, floppy hat is the next best thing. These are shields against the relatively low-energy electromagnetic radiations in sunlight. For the higher-energy electromagnetic radiations, higher up on the spectrum, heavier shielding is required.

Although a good sheet of typing paper is sufficient protection from external alpha particles and a tin can will stop beta rays, gamma rays and X-rays require special shielding. Many materials are used for industrial shielding in diagnostic X-ray facilities, nuclear power reactors, and nuclear waste repositories. A pool of water 30 feet (9 m) deep makes an excellent shield for cobalt-60 or cesium-137 radiation sources as are used for medical sterilization or food preservation. The water is conformal, and it shapes to fit perfectly around a radiation source, regardless of its configuration or surface quality. The source may be removed from the shield material without dealing with shielded doors or mechanisms. It is simply lifted out of the water with remote controls and conducted to its application point, which is usually a lead-shielded facility. The water is transparent, so the sources can be counted and inventoried in place as they are in the shield and visible. The water is a natural coolant, preventing the high-radiation gamma sources from building up heat to the melting point, as they tend to do in a vacuum or even in air. Hundreds of curies of cobalt-60, or any industrial gamma-ray source, are undetectable with sensitive instrumentation at the top of a column of shielding water. The shielding property of water is 100 percent effective.

Concrete and structural steel are common shielding materials. Concrete may be specially formulated for gamma shielding by the inclusion of steel chunks, increasing the density and therefore the efficiency of the shield. Rock and earth are used for shielding. Much low-level radioactive waste, in fact, has been simply buried under at least three feet (one m) of dirt in remote, uninhabited, federally owned locations. Current long-term plans for high-level radioactive wastes from the nuclear power industry are to store it in caverns dug out of subterranean rock formations.

Exotic shielding materials, used for deep-space probes subject to solar flares and high-energy, interstellar cosmic rays, have used tantalum and even gold to stop gamma rays with a minimum use of mass.

Depleted uranium-238, a by-product of the nuclear fuels industry, is even used for gamma shielding in special applications. Although uranium-238 is itself radioactive and therefore contributes radiation noise, it is an extremely dense material and provides five times the shielding efficiency of lead.

The most frequently used shielding material remains the first one ever used—lead. The only applications in which lead cannot be used for gamma-ray shielding are in extremely sensitive radiation-counting operations, such as carbon-14 dating or the counting of exceedingly rare radiation particles. Under these conditions, radiation noise must be eliminated. Lead does eliminate all external noise, but natural, recently mined lead, which is a mixture of lead-204, -206, -207, and -208, has a contaminant of radioactive lead-210, emitting a 0.0465 MeV gamma ray. Lead-210 has a half-life of 22.3 years. Lead is self-shielding, in that the vast majority of the gamma rays from lead-210 never escape the volume of the lead. Gamma rays do, however, stream off the surface of the lead, and any shaped shield made of this or any other material must have a surface. For low-noise radiation-counting operations, old lead, mined more than 300 years ago, must be obtained. Many counting applications use shields made of old, preatomic steel parts, such as from obsolete naval gun components. Steel manufactured after the development of large-scale nuclear fission invariably contains contamination quantities of radioactive cobalt-60.

The use of lead shielding was first made official with standards set by the Second International Congress of Radiology, in Stockholm, Sweden,

Shielding

© Infobase Learning

A single thickness of paper will shield against alpha particles α, whereas a thin sheet of aluminum will stop beta rays β. To stop gamma rays requires a great deal of lead. Neutrons (n) are slowed to thermal speed by collisions with hydrogen nuclei at room terperature.

in 1928. These standards for lead shielding for X-ray rooms are still followed, giving a minimum thickness of lead depending on the voltage applied to the X-ray tube. For the smallest, least powerful X-ray tubes using 75 kilovolts, a lead thickness of 0.039 inches (one mm) is specified. A square foot of this material weighs 2.5 pounds (1.1 kg). A full body X-ray requires 300 kilovolt X-rays, and to ensure the safety of X-ray technicians working with this equipment requires 0.354 inches (9.0 mm) of lead, weighing 24 pounds (11 kg) per square foot (929 cm²). For the most powerful industrial X-ray machines in use, at 900 kilovolts, 2.0 inches (51.0 mm) of lead is required. It weighs 120 pounds per square foot (586 kg/m²).

Finding the proper thickness of lead shielding for gamma-ray shielding is more complicated, as the working distance from the radiation source, the energy of the rays, and the time to be spent with the source per day must all be considered. The National Bureau of Standards, using newly formulated calculation methods, issued recommendations for lead shielding thicknesses, published as document NBS-1003 in 1952. The National Bureau of Standards, a non-regulatory arm of the U.S. Department of Commerce in Gaithersburg, Maryland, was reformed into the National Institute of Standards and Technology in 1988, but its original gamma-shielding specifications still stand.

An example of a shield recommendation is for handling 500 millicuries (1.85×10^{10} Bq) of a nuclide decaying with a 1.8 MeV gamma ray. The minimum working distance is 20 inches (50 cm), and the time of exposure is four hours per day. For perfect safety of the worker under these conditions, 3.93 inches (9.98 cm) of lead shielding is required. To work with a 100 curie (3.7×10^{12} Bq) source of 1.5 MeV gamma rays, eight hours a day at a distance of three feet (one m), would require 6.40 inches (16.25 cm) of lead shielding.

Workers in radiation industries, such as radiology or nuclear power, are now protected from injury or health degradation by all three dimensions of radiation avoidance. Shielding is a fixed factor that is built into the facilities, and distance from a radioactive source and time spent in a radiation field are carefully regulated. Indications of a link between health and radiation exposure were first noticed in X-ray technicians, and by 1911 alarms were raised, even though the first malpractice lawsuit award for X-ray burns was in 1899. Since then, increasing awareness and study has resulted in a greater attention to radiation protection by limiting exposure

THE TRAGEDY OF THE JAPANESE FISHERMEN

On March 1, 1954, at Bikini Atoll in the Marshall Islands of the South Pacific Ocean, the improved Ivy Mike hydrogen bomb was tested. The yield was hard to predict in this new design, but it was expected to be from four to six megatons. The atomic bomb that had wiped out Hiroshima, Japan, in 1945 yielded 16 kilotons, as a comparison. This was the first in a series of bomb tests named Castle, and this one was Castle Bravo.

Trying to predict where the radioactive fallout from the Castle tests would land, project scientists mapped a rectangular area around Bikini and issued a vague warning over broadcast radio channels for fishing vessels to stay out of this area while they were testing. In the area was a Japanese fishing boat, the *Daigo Fukuryū Maru* (Lucky Dragon 5). The 23-man crew was following the tuna, and their fishing expedition led them to a patch of ocean about 20 miles (32 km) outside the projected fallout zone, northeast of Bikini.

At 6:45 a.m. local time, the weapon was detonated aboveground. Within one second, it formed a fireball nearly 4.5 miles (roughly seven km) wide and left a crater in Bikini Atoll 6,500 feet (2,000 m) in diameter. In one minute, it formed a mushroom-shaped cloud of fine ash 47,000 feet (14,326 m) high, drifting east in the prevailing wind. Ten minutes later, the surrounding islands of Rongerik, Rongelap, and Utirik were heavily contaminated with radioactivity from the explosion. The bomb surprised its designers, yielding 15 megatons of blast and making it the largest explosion ever set off by the United States.

The crew of the *Daigo Fukuryū Maru* could not help but notice the blinding flash in the sky to the east, and six minutes later the shock wave hit the boat, sounding like a *pikodon,* Japanese for an atomic bomb explosion. As they stood on the deck, speculating, a fine, gray ash started to fall like snow. It snowed for almost three hours, and the deck, the rigging, and everyone on the boat became thickly covered with the ashes of Bikini. They had no way of knowing that it was highly radioactive. As they carefully recovered their nets and prepared to leave the area, everyone aboard started to suffer nausea, pain in the eyeballs, and skin inflammation, classic symptoms of acute radiation poisoning.

They managed to run to their home port of Yaizu in the Shizuoka prefecture of Japan and report their strange, increasingly painful ailments to doctors. Newspaper reporters picked up the news and immediately made the connection to the hydrogen

(continues)

(continued) _____

bomb test on Bikini Atoll, and soon the two worst affected crewmen were transferred to the Tokyo University Hospital for examination and treatment. Kuboyama Aikichi, the radio operator, died on September 28, 1954, after numerous blood transfusions.

The entire population of Japan was outraged by the hydrogen bomb–induced death of the man; the tuna market collapsed from fear of radioactive contamination of the fish; and a new type of Japanese horror movie, *Godzilla,* was filmed, all a result of the unpredicted size of Castle Bravo. American scientists were sympathetic, and reparations were paid to the crewmen, the widow, and the government of Japan, but they pointed out that if the crewmen had only bathed themselves after being covered with ashes, there would have been no problem. In retrospect, the fishermen should have thrown their contaminated clothing overboard and been hosed down with seawater. Washing the radioactive ashes from their bodies and the deck of the boat would have cut the exposure time to a minimum and would have maximized the distance between the crew and the radioactive sources, and there would have been no noticeable effect. Having radiation sources against the skin for 14 days was the worst possible condition resulting from the fallout.

The fishermen's boat still exists and may be seen at the Metropolitan *Daigo Fukuryū Maru* Exhibition Hall in Tokyo, Japan.

time, increasing the distance to a source, and putting a shield between the worker and the radiation.

BREATHING, EATING, AND TOUCHING

On March 14, 1954, a Japanese fishing vessel, *Daigo Fukuryū Maru,* came to port in Yaizu, Japan, to unload tuna and seek medical attention for the entire crew, who had unfortunately suffered a large radiation exposure from the aboveground test of Castle Bravo, a new hydrogen bomb tested by the United States. The bomb was of unexpected power, and dust from its disintegration and the activation of dirt underneath it rained down on the boat and its crew.

A young biophysics professor at the city university had read about the sick crewmen and wondered about a point that the medical team had missed. He was concerned with the fish that had been unloaded from the boat. If the men were contaminated with fallout, then the fish could be contaminated as well. He traced the shipment of tuna to the Osaka Central Market and found them easily using a Geiger-Müller counter. They were easily identified. They caused the headphones on his radiation instrument to click at an alarming 60,000 counts per minute, making an almost continuous buzzing sound instead of the occasional click that one hears in normal background counting.

The professor notified city officials, and they quickly found that loose scales and paper wrappings from fish already sold were heavily contaminated with radioactive dust. A hundred people had probably already eaten contaminated tuna. A fish scale involved in hydrogen bomb fallout looks the same as any other fish scale, until a Geiger-Müller probe is waved over it and the invisible radiation contamination becomes evident. The evening newspaper carried the story, and widespread panic broke out within minutes. The Japanese were unusually sensitive to radiation-induced sickness after the bombings of Hiroshima and Nagasaki, and the invisible, undetectable nature of the radiation only intensified their fears.

People in Japan stopped buying fish. Fish dealers had trouble convincing customers that their fish were not radioactive, and the Misaki market was closed on March 19. The hysteria spread quickly to Yokahama and then to Tokyo, the largest city in Japan. The great Tokyo Central Wholesale Market closed for the first time since the cholera epidemic of 1935. It was reported that fish had been eliminated from the emperor's diet, and that finished it. The tuna industry in Japan crashed.

A panic by fish buyers in Japan at the thought of eating radiation-contaminated tuna was not unreasonable. It is one thing to be injured by fallout dust on the skin. Although the alpha and beta radiations cannot penetrate and cause harm to the internal body organs, the gamma rays from the fission debris certainly can penetrate the body at any contamination point and exit through the other side, leaving a trail of damaging ionization but not necessarily wreaking all possible destruction. The intruding gamma ray has energy left over after it leaves the body. Actually eating food contaminated with pieces of fission products, however, is another matter. In this case, the danger from alpha, beta, and gamma radiation damage to internal organs is extreme. Alpha particles are weak only in their ability to penetrate matter. In all other aspects, they are the deadliest of radiation types. When one ingests an alpha particle source, the entire energy of the alpha decay is deposited inside the body. One

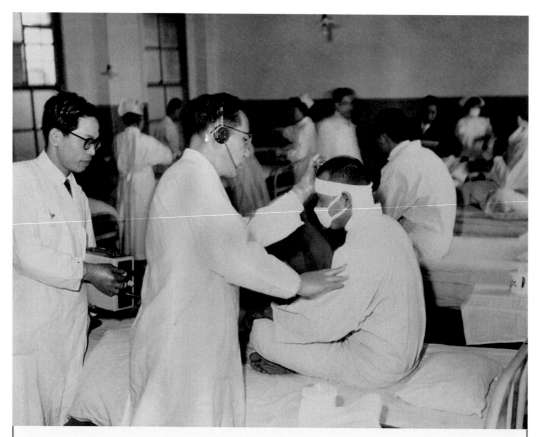

The treatment of a Japanese fisherman exposed to fallout after the U.S. nuclear test at Bikini Atoll in 1954 *(© Bettmann/CORBIS)*

hundred percent of the ionizing power of the alpha particle is left in the tissue when an alpha emitter is consumed. The same rule applies to beta rays, although they have a fraction of the destructive power of alpha particles. Gamma rays, while less able to deposit entire energy packets in living tissue, are still dangerous, and they definitely contribute to the hazard of ingesting radiation sources.

The human body treats an ingested radionuclide like any other food item taken through the mouth. The long chemical plant of the digestive system breaks it down into absorbable compounds and then tries to find a use for each chemical, no matter how unusual. If no use can be found for a substance, then it usually passes through the system, but not before radioactive components have attacked the lining of the entire system. The digestive system, which constantly rebuilds its liner due to the mechanical stresses of continuously moving food downstream, is one of the most sensitive tissues in the body to radiation damage. As the cells divide to make new lining material, they are wide open to ionization distress on the twinning DNA strands. Thus, the gastrointestinal tract can be negatively affected by ingested radioactive material, such as fallout-contaminated fish.

A common fission product radionuclide is iodine-131, and iodine-131 was probably in the tuna shipped from the Misaki market on March 14, 1954. The Castle Bravo hydrogen bomb used a plutonium-burning fission bomb as the initiator, and its fission products included iodine-131. The tuna in the last net of fish hauled aboard the *Daigo Fukuryū Maru* had actively ingested the fallout dust as it settled in the water and had already begun to metabolize such nuclides as iodine-131.

The body's chemical-sorting system finds an immediate need for iodine and is insensitive to the subtle nuclear differences in the various nuclides of iodine. The naturally occurring nuclide of iodine is iodine-127, and the digestive system usually scavenges it from fish or table salt, where iodine is an artificially added nutrient. The radioactive iodine-131 is sent directly to the thyroid gland, where it is naturally concentrated. As a result of this specialized use of iodine, when presented with radioactive iodine, the thyroid gland suffers greatly and in significant doses is likely to develop cancer.

Strontium-90, a radioactive component of nuclear fallout and fission waste, is readily absorbed in minute quantities by the human body. Although the body has no particular use for strontium, which chemically resembles calcium, it is conducted through the bloodstream to the bones. Bones, being living tissue, must constantly maintain structural integrity, and one of their main building blocks is calcium. When strontium-90 is concentrated in the bones, it most affects the bone marrow, where blood cells are constantly being manufactured. Bone rebuilding is a very slow process, but blood cells are built quickly and in great quantity and thus are very sensitive to radiation poisoning. Leukemia can result from ingesting significant quantities of strontium-90. Iodine-131 has a half-life of 8.04 days and loses its danger potential after 10 half-lives, or 80 days. Strontium-90 has a half-life of 29 years and remains dangerous over several half-lives.

Another way of bringing radioactive material into the body is to breathe it, and this is why the men aboard the *Daigo Fukuryū Maru* should not only have washed themselves and discarded their clothing but washed

Workers in full dress clean up radioactive dust at the Commissariat à l'énergie atomique in Fontenay-aux-Roses, just outside Paris, France. *(Patrick Landmann/Photo Researchers, Inc.)*

down the entire boat. Not doing so after being subjected to radioactive dust from the Castle Bravo detonation caused them to breathe the dust for 14 days, until they reached their home port in Japan. Breathing does not have quite the potential for global positioning of radionuclides in specific body organs, but it does subject the very sensitive tissues of the lungs to a wide range of dangerous radiation sources. A radioactive dust particle can stick in the lungs, where 100 percent of its radiation emissions interact with the body.

Emanations from a bit of radioactive matter, such as a unit of dust, go in all directions on a purely random schedule, with no preferred angle of launch. An imaginary surface of equal radiation intensity around a radioactive dust particle is therefore spherical. If one stands a distance from such a particle, the sphere of constant intensity has a radius of the distance. The farther away from the particle a person is, the larger the sphere surface, meaning that the unchanging amount of radiation is spread out over a larger area. The farther one moves from the source, the longer is the radius of the imaginary sphere, and as the sphere grows in size, a lesser percentage of its surface area covers a person standing at the radius. For this reason, the radiation dose rate decreases with distance according to the square of the radius, or the surface area of the sphere. When the radiation source is trapped in the lungs, a person is subjected to the entire surface area of that imaginary sphere of constant intensity. Every particle of radiation that is launched from the source hits the inside of the lung.

PRACTICAL RADIATION MONITORING

If a radiation industry or nuclear research worker happens to step on a drop of radioactive liquid that accidentally landed on the floor, then the remainder of the worker's steps that day can be traced, from the stroll out to the bus stop going home, on the bus, off the bus, across the street, and to the door of the domicile, at least until the work boots are removed. Everywhere the boot stepped after being contaminated by the drop of liquid, a small portion of the droplet was deposited, and if the activity of that droplet was significant, then those steps can be followed using radiation-detection equipment.

There are different levels of radiation-detection equipment. In a nuclear reactor, there are fixed instruments, bolted in permanently, to monitor the power level of the fission process. In the hallways and portals of a nuclear power plant, there are fixed instruments used to monitor employees leaving and arriving, searching constantly for unusual levels of radiation. It was just such a portal monitor at a nuclear power plant in Forsmark, Sweden, that first detected the Chernobyl reactor disaster in 1986. Dust from Chernobyl had traveled hundreds of miles in the upper atmosphere, fallen out over Sweden, and then was tracked into the radiation detection portal on a worker's shoes. Before the portal alarm sounded, there was not a hint that anything untoward had happened in Chernobyl.

Enormous radiation portals, weighing several tons, are now used to monitor incoming ships' cargo and entire tractor trailers for unusual shipments of radioactive materials. These instruments are so sensitive that a truckload of cat litter, a clay mineral containing very small amounts of uranium and thorium, will set off an alarm as if nuclear weapons were being moved surreptitiously. In a radiation research laboratory, there are instruments of all types bolted into equipment racks and sitting in fixed locations on benches. The subtype of radiation counter that is used to find radiation leaks, spills, contaminations, or radioactive minerals in the field is the handheld detector.

The most commonly used handheld detector is the Geiger-Müller counter, often shortened to simply Geiger counter for the sake of brevity. The instrument consists of a small, metal box with a handle built into the top. The sensitive piece, a small, hand-sized Geiger probe, is usually connected back to the box with a curled cable, but the probe may also be built into the bottom of the box. The Geiger counter is relatively inexpensive, is dependable and rugged, and can detect very small amounts of different types of radiation. It is, however, less than 1 percent efficient. If 100 gamma rays penetrate the Geiger probe, then one of the gamma rays has a chance of being counted. The Geiger counter gives no indication of the energy of a detected ray, and the nature of detected radiation can be derived only from the amount of filtering material covering the Geiger tube. For alpha particles, an extremely thin covering, transparent to alphas, is used, but in this configuration beta and gamma rays are also counted. Similarly, for beta rays a thin window of aluminum is used, and this arrangement rejects alpha particles but also counts gamma rays. With a steel beta shield over the Geiger tube, the counter will register only gamma rays, filtering out all other types of radiation. Beta and alpha ray counting in a Geiger counter is much more efficient than is gamma counting, and this gives a skewed impression of the gamma radiation intensity compared with alpha and beta in a radiation survey.

The Geiger counter probe can sample only those radiation particles that actually hit and penetrate the Geiger tube inside the probe. The presented area of a Geiger tube in a handheld instrument, a cylinder typically one inch (2.5 cm) in diameter by three inches (7.6 cm) long, does not provide a very wide area of detection. For this reason, a point source of radiation, such as a bit of cobalt-60 dust or a droplet of water containing cobalt-60, although extremely active, can easily be missed with a Geiger counter. The sphere of constant activity from a radiation source, at a distance, casts a minor shadow against a Geiger probe. The probe must be held close to the radioactive fragment to detect an unusual radiation level, practically touching the source. Searching for the next radioactive footstep of the hypothetical radiation worker that has stepped on a droplet can be a laborious, on-the-knees job, covering every square inch of floor.

An exception to this rule occurs when the radiation source is not a point, but instead a surface. An example of a surface source is an uncovered geologic formation of uranium ore, absent an overburden of radiation-shielding topsoil and vegetation, as is found in the desert regions of the American Southwest. With an entire landscape emitting radiation, the square-of-the-distance rule for radiation intensity falloff no longer applies, and gamma rays can be counted from an aircraft flyover using Geiger or scintillation counters. Uranium ore surveys have been carried out with portable radiation detectors mounted in jeeps, helicopters, and single-engine airplanes. Detailed maps of ground radiation, showing iso-radiation contours, have been made of uranium fields, oil well penetrations, and contaminated plant and laboratory sites using aerial surveys.

An improvement on the original Geiger counter was the addition of a ratemeter, which is an analog readout, showing a scale of radiation intensity with a needle pointing to the current value. The usual jerky nature of the radiation count rate is dampened by electronically averaging the count rate over a specified span of time. This feature causes the indicator

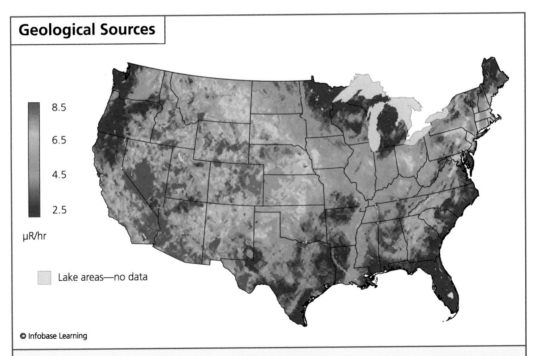

Geological Sources

8.5
6.5
4.5
2.5

μR/hr

Lake areas—no data

© Infobase Learning

A map of the continental United States, showing the radiation flux originating in surface minerals, expressed in micro-roentgens per hour; this map was compiled from aerial surveys using scintillation counters mounted in aircraft. Some of the highest ground radiation is shown in Nevada, where 20 years of aboveground nuclear weapons testing has added to the natural radiation from minerals. Water is an ideal radiation shield, so no radiation registers over lakes or rivers.

needle to move smoothly as the count rate can change, but it does result in a delayed reaction to a changing count rate. In Geiger counters currently built, the analog meter is often replaced with a digital readout, but the principle and the response limitation are the same.

Handheld Geiger counters, scintillation counters, and ionization chambers are still available with rems or sieverts per hour on the rate-meters for use in health physics radiation surveys, but most professional instruments now indicate counts per minute or counts per hour, the only measures that a radiation counter can accurately give. A radiation counter measures only the number of rays successfully ionizing the detector tube per unit time. It is not aware of the energy of each ray, which organ of the body is of interest, or how many rays are hitting the person who is holding the instrument, so it can give only a rough estimate, based on averages, of the effect the radiation is having on a living human being, which is the meaning of the rem or sievert.

In radiation detection portals or self-test stations in nuclear power plants a radiation detector is not looking for the number of rems or sieverts per hour. It is looking for the onset of a difference in the number of counts per minute, and it is set to alarm if the count rate suddenly increases by a set amount. A radiation worker leaving the laboratory may test both hands for radiation picked up in the lab, by holding one palm and then the other in a Geiger probe. As is the case of a whole-body portal, the self-test station will set off alarms if it detects a positive change in the count rate, and the rem rate is not considered.

7 Medical Treatment of Radiation Poisoning

Much was learned in the half century after the mass radiation casualties of the atomic bombing of Japan and radiation accident experiences in the early years of nuclear technology. Before there were atomic bombs and gigawatt-producing nuclear reactors, there were only a few cases of radiation sickness, usually involving radium-226 or X-rays. Medical science, with few examples to study, moved slowly toward understanding the effects of unusual radiation exposures and how to treat them. Initial studies of atomic bomb survivors and laboratory and worker irradiation incidents eventually led to symptom recognition, followed by more detailed prognosis, and finally the formulation of treatment options for specific radionuclide contaminations.

In the 21st century, the danger of radiation poisoning has gained renewed interest in the military and medical communities. Although the threat of a thermonuclear exchange between superpowers may have dissipated, there are now different concerns for mass radiation poisonings. Terrorist activity, once a small component of national security, has moved to primary importance, and a few new nuclear attack scenarios are now feared. Terrorists could attack anywhere using low-level atomic bombs detonated at or near ground level. More likely, terrorists could use a dirty bomb, or radiation dispersal device.

A dirty bomb is a radiological weapon intended to spread a portion of a dangerous radioactive nuclide by use of a chemical explosive. Examples of usable dirty bomb active components are cobalt-60 and cesium-137, dispersed into a crowded city using any of a number of available high explosives.

Adding to the nuclear fears of the new century is a continuing and increasing use of nuclear electrical power worldwide. While nuclear power has a fine safety record, there are instances, such as Chernobyl in 1986, where a civilian power station breakdown sterilized an inhabited area as well as contaminated an entire subcontinent. The earthquake and tsunami disaster in northeastern Japan in 2011 spread fission products from three melted reactor cores over hundreds of square miles of farmland, causing heightened concern for radiation contamination of agricultural products. Spills of transported fission waste products or even terrorist-produced accidents involving fission waste, while made unlikely by designed safety measures, are not impossible.

Governmental concern for such accidents or enemy actions has resulted in much work toward medical treatment of radiation sickness victims. This chapter gives an outline of the procedures that now exist for the treatment and prevention of radiation injuries. There is a strong psychological component of radiation injury, and an illuminating example of this symptom is detailed in the sidebar on pages 121–122, which recounts an odd incident in Brazil in 1987. What started out as a small nuclear accident turned into a mass panic, and eventually 100,000 people had to be checked for radiation injury.

RADIATION EXPOSURE SYMPTOMS AND DIAGNOSIS

In the event of radiological terrorism or a nuclear power accident of the magnitude of Chernobyl, first responders are being trained to quickly diagnose and triage victims, separating radiation-exposure injuries from more conventional injuries and then sorting radiological patients according to dose received. Specific guidelines have been formulated for rapid radiological triage.

The first question to ask a patient is, "How long since the explosion until you started vomiting?" As has been observed in all cases of extreme, acute radiation exposure, a dosing greater than 50 rads (0.5 Gy) will cause nausea, with the onset delay inversely proportional to the dosage. Patients are immediately broken into two groups: those who started vomiting with a delay of four hours or less and those that vomited after four hours. Those with the early onset require immediate evaluation. Those with the late onset are referred for later evaluation, after 14 to 72 hours. A patient expe-

THE STRANGE CASE OF THE GOIÂNIA ACCIDENT

Goiânia is the capital of the Brazilian state of Goiás. It is home to a million people, a university, many tall buildings, and a busy airport. In 1971, a new cancer treatment clinic was opened in downtown Goiânia, using the latest in radiation therapy, a machine weighing several tons that could deliver a precisely controlled dose of gamma rays to a tumor. The source of gamma rays was 2,000 curies (74 TBq) of cesium-137, in the form of cesium chloride mounted in a canister made of lead and steel, fitted with a gamma window made of iridium.

By 1978, a cesium-137 gamma ray machine was considered obsolete, with most cancer treatment being upgraded to cobalt-60 as the radiation source. In December 1985, the clinic and its newer cobalt-60 machine were moved to a better location. The old hospital was torn down, but the clinic and its old cesium-137 machine were tied up in litigation, and the site was abandoned.

On September 13, 1987, two scavengers entered the abandoned clinic building, removing anything that could be carried. The only thing remaining in the shell of the building was the hulking gamma machine, too heavy to move. The lead cylinder in the gamma machine looked valuable, and they managed to pry it free of the machine. One

(continues)

A woman with gamma radiation burns from the Goiânia cesium-137 release incident in Brazil
(© Karen Kasmauski/Science Faction/Corbis)

(continued)

of the scavengers took it home. Then, the two scavengers spent four nights working on it with a hammer, and by September 16 they succeeded in breaking through the iridium window, getting dosed with gamma rays all the time at a rate of about 314 rads per hour (3.14 Gy per hour). The two suffered severe radiation burns, and one later had to have an arm amputated at the elbow, but at the time both were intrigued by the blue glow coming from the interior of the lead cylinder.

Knowing they had something valuable, the two scavengers went to the local junk-yard on September 17 to sell it. On September 18, the junk dealer finally agreed to their price of 1,600 cruzados, or about $25. He planned to make a ring out of the glowing material for his wife, so he took it home and hid it in her closet.

On September 21, the junk dealer's wife started to notice that everyone in the house was sick and suffering from diarrhea. Thinking it had to be the soft drinks they were consuming, she took a sample to the local authorities for testing. Two days later, a couple of workers from the junkyard managed to break open the inner cylinder, spilling cesium-137 all over. Everywhere they went, contamination resulted. They would die 39 days later of radiation sickness in a Rio de Janeiro hospital. The junk dealer's brother took some of the powder home, spreading it out on the floor where his six-year-old daughter was eating. She spread it over her body and showed the glow to her mother. People came to see it, spreading contamination over the neighborhood.

On September 29, a visiting medical physicist finally waved a scintillation counter over the powder and found it alarmingly radioactive. That evening, the government of Brazil began the accident response. Over 100,000 people had to be examined for radiation effects, 244 had been seriously contaminated, and 46 required immediate hospitalization. Six people died within weeks, including the junk dealer's wife and his brother's young daughter. In the future, it is expected that 20 percent of those who ingested the cesium-137 dust will contract leukemia.

In the massive cleanup operation, entire neighborhoods had to be demolished, with the topsoil removed and taken to a burial site. Some houses were saved by cleaning the floors, walls, and even the roofs, and two had to have the roofs removed. The population of Goiás seemed to panic as the newspaper accounts of deaths and cleanup squads fanned fears, and thousands showed up at emergency rooms with psychosomatic symptoms. Hospitals and clinics were overwhelmed, and all manner of state commerce froze in place for two months.

With enormous expense and effort, decontamination teams were able to recover all but 13 percent of the cesium-137. By now, the unrecovered material has an activity of about 108 curies (4 TBq), and its location is unknown.

riencing nausea within one hour of exposure will require extensive and prolonged medical intervention, and fatal outcome is very likely. The dose received in this case is between 500 and 1,100 rads (five and 11 Gy). The onset of vomiting and other medical tests are biodosimetry, a means of evaluating the integral dose received by a patient not wearing a film badge or dosimeter.

The second rough triage observation is a complete blood count. A blood sample is taken, and the lymphocyte density is evaluated. Fewer than 1,000 lymphocytes per cubic centimeter suggest moderate exposure. Fewer than 500 lymphocytes per cubic centimeter suggest severe exposure. A normal lymphocyte count eight to 12 hours after exposure is 2,500 lymphocytes per cubic centimeter. With an exposure of 100 to 500 rads (one to five Gy), the count drops to 1,700 to 2,500. With 500 to 800 rads (five to nine Gy), it drops to 1,200 to 1,700. Fewer than 1,000 lymphocytes per cubic centimeter of blood indicate a total exposure of more than 1,000 rads (10 Gy). The lymphocyte count in this range is as accurate an indicator of dosage as any dosimeter worn in a shirt pocket, and this measurement gives an accurate reading as to the level of treatment a patient will require, as well as the probability of survival.

Observation of a patient's skin over a period of three weeks also gives an accurate measure of an external radiation dose. The beginning of epilation, or the loss of hair, indicates a dose of about 300 rads (three Gy). The entire hair shaft falls out, often in clumps. Erythema and desquamation, or the reddening of skin and then the scaling off of the outer layers of skin, have been experienced by anyone with sunburn. In the case of gamma or beta burn, erythema after two to three weeks means a dose of 600 rads (six Gy). Dry desquamation in the same period indicates a dose of 1,000 to 1,500 rads (10 to 15 Gy), and wet desquamation means that the patient externally absorbed 2,000 to 5,000 rads (20 to 50 Gy). A dose of more than 5,000 rads (50 Gy) leads to overt radionecrosis and ulceration after four weeks.

CLINICAL PHARMACOLOGY AND THERAPIES

The first medical treatments for radiation injuries were formulated for laboratory workers who had somehow ingested radionuclides, starting with the ancestral beginning of radioactive material research, radium-266, discovered in 1898. Soon after radium was discovered as a rare earth in uranium ore, it was discovered that radioactive materials can enter the body through a number of openings. Cathartics were used to move material quickly through the digestive tract and minimize the internal exposure time and therefore the dosage from radioactive substance in the bowel. A biscodyl or phosphate soda enema will empty the colon in a few minutes and is still given primary consideration for the treatment of swallowing radioactive material. In the now rare case of ingesting radium, magnesium sulfate is quite effective as it produces insoluble compounds with the radionuclide and keeps it from entering the bloodstream. In the early days of radiation laboratory work, it became clear that food and drink should not be eaten anywhere near radioactive material, as the slightest contamination will eventually wind up on the hands, and what is on the hands will wind up in the mouth. Since then, several pharmaceuticals have been developed specifically for the treatment of radiation exposure, inhalation, and ingestion of specific radioactive substances. One promising medicine, amifostine, has even proven to prevent cell damage by radiation, but it must be administered before radiation is encountered.

DECONTAMINATION

In any nuclear emergency involving the spread of radioactive debris, decontamination is of secondary importance. The immediate triage and treatment of the injured is primary, but the decontamination is necessary to prevent the development of further casualties as more people are exposed to radiation. Everything that has been sprinkled with radioactive material, including the victims, must be cleansed, and the contaminated material must be consolidated and disposed of under shielding.

The Goiânia cesium-137 accident in Brazil, as detailed in the sidebar on pages 121–122, is an example of a learning experience in radiation cleanup. Industrial vacuum cleaners wielded by workers dressed in protective coveralls proved the most effective way to decontaminate streets, yards, and entire buildings. The clothing and respirators worn by cleanup personnel are not radiation-proof clothing. They are intended to keep radioactive dust from collecting on the skin and in the lungs of the workers and include booties to cover the feet and gloves to cover the hands. The work is hot and uncomfortable in radiation-protective gear, but at the end of the shift the garments are carefully removed, not touching the outside surfaces, and discarded, without the worker having any direct contact with radioactive material. Gamma rays go right

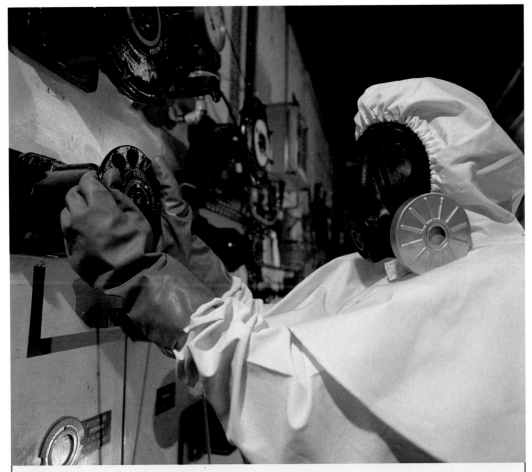

A worker wearing a radiation suit with breathing apparatus at a nuclear power station in England. *(Steve Allen/Photo Researchers, Inc.)*

through the clothing and remain dangerous, but the dosage is carefully monitored using individual dosimeters, and there is no residual, lingering radiation effect as long as the worker removed his suit and respirator correctly. A primary goal of decontamination is to not spread radioactive material any farther than it is already dispersed, and this first means that the workers cannot be contaminated and that all equipment used is tightly controlled.

In the Goiânia case, Prussian blue was administered for cesium-137 chelation for the victims, and it was used on walls and floors in buildings to round up invisible quantities of the radioactive nuclide. Potassium alum dissolved in hydrochloric acid was used to pick up the cesium on bare clay, cement sidewalks, streets, driveways, topsoil, and roofs. Waxed floors and tables were cleaned with organic solvents, followed by the potassium alum solution. In houses that required demolition, personal objects that had no contamination, such as photographs or jewelry, were placed in plastic bags and removed, in hopes of reducing the psychological impact of the most severe cases of decontamination.

In the hypothetical case of a mass attack by a radiation dispersal device or a nuclear weapon, decontamination of victims begins with the removal of clothing. Ninety percent of all bomb-delivered radioactive contamination is disposed of simply by taking off clothes, being careful not to cross-contaminate the hands and arms by touching the outer surfaces.

Checking the radiation level of a container in an air force major accident response exercise; the airmen are fully suited up for protection against radioactive dust. (*U.S. Air Force; photo by Nan Wylie*)

The patient is then moved to a shower, where the remaining 10 percent of surface contamination is removed using soap.

Hair collects an unusually high percentage of radioactive debris and must be washed with any commercial shampoo. There should be no conditioner used, as conditioners bind material to the hair protein. If a handheld radiation counter shows that the hair is still radioactive after vigorous washing, then it must be clipped off, bagged, and removed. Cleaning solutions that have been tried and recommended for further skin decontaminations are normal saline, providone iodine in water, and 3 percent hexachlorophene in detergent and water.

The philosophy of decontamination at its core is based on the three principles of radiation avoidance: minimize time, maximize distance, and use shielding. By removing radioactive dust from human skin and surfaces that may be touched, the distance from the radioactivity is maximized, and the time spent near it is minimized. Contaminated materials are always buried, putting a shield of earth between the radioactivity and mankind.

LONG-TERM EFFECTS

Although the long-term effects of large-scale radiation contamination have long been studied, there is still no certain answer to what will happen to a population because of extended exposure. It can be stated that chronic radiation exposure of sufficient intensity will likely cause certain types of cancer, but there are so many possible alternate causes of cancer and so many variables it is difficult to pin down the cause. Cancers that are known to be caused by the ingestion of common fission products are thyroid cancer, caused by iodine-131, and leukemia or bone cancer, caused by strontium-90. Because iodine has a short half-life of eight days, it is not considered a long-term health risk. After 10 half-lives, or 80 days, the intensity of iodine-131 radiation has fallen by a factor of more than 1,000.

Cesium-137, a fission product with a half-life of 30 years, is dangerous and long-lived, but it has a biological half-life of only 140 days, which is to say that after 140 days half of the ingested cesium-137 will have been excreted from the body, even with no drug therapy. Cesium is not commonly used by the human body for any specific function and therefore does not concentrate in a sensitive place, but usually is dispersed in the muscles.

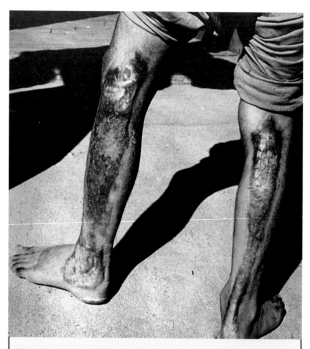

The damaging effects of heat and radiation on a man's legs from the nuclear attack on Hiroshima in 1945; these burns were caused by direct exposure to the extreme radiation pulse from the weapon and not by prolonged exposure to radiation. *(National Archives)*

Strontium-90, however, chemically resembles calcium and therefore goes straight to the bones. It has a half-life of 28 years and is safe after 280 years. There is a slow turnaround for calcium or strontium in the bones, so it has a 10-year biological half-life, concentrated near the marrow and the epiphyseal plates. In these locations for extended times the beta rays from strontium-90 can do a great deal of damage. Of the hundreds of radionuclides produced by nuclear fission, strontium-90 is probably the most dangerous in the long term.

Beginning in 1945 and ending in 1963, two deployed nuclear weapons and hundreds of experimental nuclear weapons were detonated aboveground, spreading atomized fission products over the entire globe. These radionuclides, including cesium-137 and strontium-90, are still in the process of decay. Although the radioactivity has fallen exponentially since the Nuclear Test Ban Treaty in 1963, radiation from nuclear tests made in the 1950s is still detectable on the ground and in the air. It is in the food and in every breath taken. In all of the nuclear tests, about 30 megacuries (1×10^{18} Bq) of cesium-137 were released.

In comparison, 2.5 megacuries (9.3×10^{16} Bq) of cesium-137 were released into the atmosphere above Europe in the Chernobyl accident of April 26, 1986. The release was not through the impulsive, high-temperature explosion of a nuclear weapon, but from first a steam explosion and then a fire, resulting in larger, heavier radioactive dust, blown by winds lower in the atmosphere. About 60 percent of the radioactive fallout from Chernobyl fell in Russia, with the remaining 40 percent spread all over Europe in a northwest to southeast swath, from Sweden to Greece.

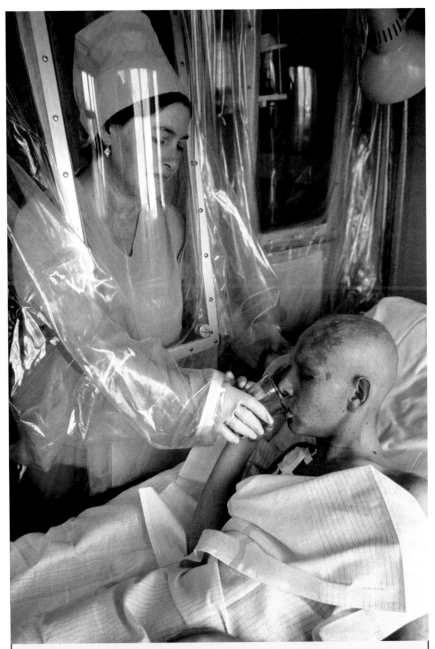

A 27-year-old cancer patient in a hospital in Moscow; the cancer was probably the result of prolonged exposure to fallout from the Chernobyl-4 reactor explosion in the Ukraine. *(Novosti/Photo Researchers, Inc.)*

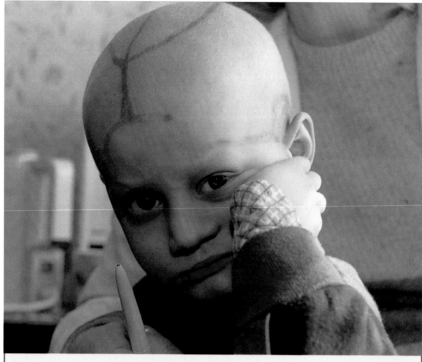

A five-year-old boy with cancer caused by the Chernobyl-4 disaster. *(AP Images)*

The total radiation release from Chernobyl is estimated at 100 megacuries $(4 \times 10^{18}\,\text{Bq})$.

The long-term radiation exposure to people living is Europe is estimated at 29 million people receiving one rem (0.01 Sv). In this population, probably 9.5 million people will contract cancer without the Chernobyl fallout over a lifetime. With the radioactive fallout and the average exposure, 3,000 more cancer deaths might occur, giving a total number of cancer deaths of 9.503 million.

Conclusion

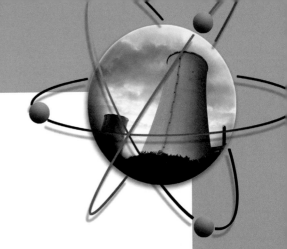

As nuclear power, nuclear medicine, and the nuclear industry in general develop further in the coming decades, many questions will arise in the scientific, engineering, and public communities. These questions will usually involve the primary safety consideration of all nuclear enterprise—radiation. With the information presented in this volume, it is hoped that many of these questions may be answered.

An example involves the ongoing debate concerning whether using a cell phone causes brain cancer. A cell phone is a handheld device used by pressing it to the side of the skull. Out of the top of the device streams electromagnetic radiation, or radio waves, at a frequency of about one gigahertz and a power level of one watt. Using a cell phone several hours of the day thus subjects the brain, which is extremely close to the antenna, to a sustained blast of radiation. People who have used cell phones day after day for years have developed brain cancer. There seems to be a correlation. Is the cell phone service provider responsible for people dying of brain cancer? Should handheld cell phone use be banned? Given information in this volume, it is possible to reach an informed conclusion. Some important points are summarized:

❋ The one gigahertz radio signal from a cell phone is indeed on the spectrum of electromagnetic radiation. It occurs on the spectrum below infrared light, which is below visible light, which is below ultraviolet light.

❋ Electromagnetic radiation is capable of ionizing atoms, but only if it is in the higher end of the ultraviolet sub-spectrum. At this point, the photons are just able to separate the electron out of its orbit in the hydrogen atom. This is the least-energy requirement for ionization. Heavier atoms and molecules require more energy. Gamma rays, near the top of the spectrum, are actually able to tear apart complex molecules, but ultraviolet is down on the scale of frequency and energy and is barely able to ionize a naked hydrogen atom. No radio wave, regardless of how high its frequency, is able to ionize anything.

❋ The mechanism of cancer development in living tissue by radiation exposure is well understood. The extremely complex DNA, or deoxyribonucleic acid, molecule can be torn by an incoming particle of ionizing radiation. DNA contains information essential to cellular repair and duplication. There is a finite probability that the modification of this molecule will cause the DNA to impart incorrect instructions to the duplication process and cause it to run wild. This is cancer.

❋ DNA is sensitive to this radiation damage only at the rare time when it is splitting in half to duplicate itself. Otherwise, the molecule always contains duplicate, redundant information, and any random ionization can be repaired by using the undamaged information as a model.

❋ Some body tissues, such as bone marrow and the gastrointestinal lining, are constantly being duplicated and are therefore unusually susceptible to radiation-caused cancer. The brain and its associated nervous system are not included in this set of tissues. The brain does not engage in cell duplication.

Collect these facts, apply them to the problem of cell phone–induced brain cancer, and a conclusion becomes evident. There is no ionization from the radiation of a cell phone. It is no more harmful to the brain than turning on a flashlight and holding it to the side of the head. Furthermore, if a cell phone emitted radiation in the gamma-ray end of the spectrum, it would not produce brain cancer. Ionizing radiation of sufficient duration and intensity could induce cancer in the skin covering the head or the bones of the skull, but it is improbable that it can cause cancer of the brain.

Radiation is dangerous and must be treated with caution in all interactions, but it is not the cause of all human ailments. Animals and humans have always been bathed in high-energy radiation, but not to the point where it was dangerous. For that reason, we evolved with no way to sense it. This was fine until recently, when through technology we introduced intense, concentrated radiation sources, and heightened vigilance is necessary to protect ourselves against this artificial threat.

1675   The English scientist Isaac Newton (1643–1727) publishes his *Hypothesis of Light,* describing light in terms of particles of matter, emitted in all directions.

1787   Abraham Bennet (1749–99) invents the gold leaf electroscope. It is still in use to detect ionizing radiation.

1789   Martin Heinrich Klaproth (1743–1817), a German chemist, discovers a metal with new and unique properties in a sample of pitchblende by a series of mistaken assumptions. He names the new element uranium, after the newly discovered planet, Uranus.

1800   Infrared radiation is discovered as an invisible component of sunlight by William Herschel (1738–1822), an astronomer with the Royal Society of London.

1801   The ultraviolet band of radiation is discovered by Johann Wilhelm Ritter (1776–1810), a German chemist and physicist, as an invisible component of sunlight beyond the violet end of the spectrum.

1864   A Scottish physicist named James Clerk Maxwell (1831–79) at King's College, London, predicts the existence of electromagnetic waves in a purely mathematical, nonexperimental exercise.

1868   Anders Jonas Ångström, a Swedish physicist, (1814–74) creates a spectrum chart of solar radiation.

1875   X-rays are discovered by Johann Hittorf (1824–1914), a German physicist who is investigating high-voltage vacuum tubes at the University of Münster.

1881   Pierre Curie (1859–1906) and his brother, Paul-Jacques Curie (1856–1941), develop a piezoelectric version of the electroscope and rename it electrometer. It is much more sensitive to electrical charge than the antique version of this instrument.

1886   Ivan Pulyui (1845–1918) of Ukraine develops a device that emits what are later known as X-rays. He names it cold light.

1887	Nikola Tesla (1856–1943), an ethnic Serb from the village of Smiljan in Croatia, develops a single-electrode X-ray tube and uses the X-rays, which are yet to be named, to photograph the bones of his foot.

At the University of Karlsruhe, Heinrich Hertz accidentally confirms the existence of Maxwell's theoretical waves, finding with further study that these electric waves behave exactly as the mathematical equations predicted they would.

1895 Wilhelm Roentgen (1845–1923), a German physicist at the University of Würzburg, makes a further discovery of Maxwellian waves, but these invisible rays are of higher energy and frequency and are capable of penetrating solid objects in ways that visible light cannot. He names them X-rays.

1896 Frenchman Henri Becquerel (1852–1908) accidentally discovers an even higher energy electromagnetic wave. These highly penetrating waves require no high-voltage electrical apparatus for production.

1898 Radium is discovered in uranium ore by Marie Skłodowska Curie (1867–1934) and her husband, Pierre Curie. An industry is created, using radium for everything from self-illumined watch dials to medical implants.

1900 Gamma rays are discovered by Paul Villard (1860–1934), a French chemist and physicist, in a study of uranium and radium.

1903 William Crookes (1832–1919), a British experimental physicist, invents the spinthariscope, using the phosphorescent properties of zinc sulfide. It is the first scintillation counter.

1905 The English physicist J. J. Thomson (1856–1940) discovers the delta ray.

1908 Ernest Rutherford (1871–1937) and Hans Geiger (1882–1945) invent the Geiger counter, a highly sensitive detector for ionizing radiation.

1909 Ernest Rutherford, the great experimentalist from New Zealand, uses spinthariscopes at the University of Manchester in England in radiation-counting experiments to discover the atomic nucleus.

1912 Victor Hess (1883–1964), an Austrian-American physicist, goes aloft in a balloon carrying radiation detection instruments to study background radiation. He discovers cosmic rays.

1917   The U.S. Radium Corporation begins operation in Orange, New Jersey, to make Undark, a luminescent paint employing radium-226 with its 4.8 MeV alpha particle and its 1,600-year half-life.

1918   The rad is developed for use in expressing quantities of X-rays used in radiation treatment of cancer tumors.

1922   The first radiation badge dosimeter is used to measure the accumulated radiation dose to X-ray technicians.

1927   The American physicist Arthur H. Compton (1892–1962) is awarded the Nobel Prize in physics for discovering that photons can collide with electrons and knock them askew.

1928   Hans Geiger (1882–1945) and Walther Müller (1905–79) improve the Geiger counter using chemical ion quenching and name it the Geiger-Müller counter.

The first company to become fully organized for the manufacture of radiation instruments, the Victoreen Instrument Company, is founded in Cleveland Heights, Ohio, by John A. Victoreen (1902–86).

The use of lead shielding is made official with standards set by the Second International Congress of Radiology in Stockholm, Sweden.

1930   The neutrino is theorized to exist by Wolfgang Pauli (1900–58), an Austrian theoretical physicist at Eidgenössische Technische Hochschule Zürich, Switzerland.

1931   Karl Guthe Jansky (1905–50) discovers radio astronomy at the Bell Telephone Laboratories. He finds that radio interference is coming from the center of the Milky Way galaxy.

1932   The first direct-reading pocket dosimeter is hand-built by Charles C. Lauritsen (1891–1968), a Danish-American experimental physicist working at Caltech in Pasadena.

1936   The first muon is detected by Carl Anderson (1905–91) at Caltech during a study of cosmic rays.

1938   The fission of uranium-235 is discovered in December by chemists Otto Hahn (1879–1968) and Fritz Strassmann (1902–80), working at the Kaiser Wilhelm Institute of Physics in Berlin, Germany.

1941   The first ionization smoke detectors, using radium-226 as an alpha particle source, are sold.

1942   At the University of Chicago, the first nuclear reactor produces power at a rate of one-half watt.

1944   Fifteen Victoreen model 247 Geiger counters deployed in Operation Peppermint at the invasion of Normandy fail to find any evidence of German radiological warfare.

At the Manhattan Project, film badge dosimeters are first used in a nonmedical situation at the Clinton Works in Oak Ridge, Tennessee.

1946   The Victoreen Model 247A special ionization chamber is used for radiation monitoring in atomic bomb tests in the South Pacific.

Karl Z. Morgan (1907–99) and Lyle B. Borst (1920–2009), physicists engaged in war research at the Clinton Works, invent a modification to the film badge dosimeter so that it will record fast-neutron exposure.

1947   Robert Hofstadter (1915–90), a professor of physics at Stanford University in California, invents the scintillation counter.

William "Willy" Higinbotham (1910–94), an American physicist, builds the first single-channel analyzer, the Model 200 Pulse counter.

1948   Sidney Liebson (1920–) substitutes halogen gas for the argon-ether mixture used in Geiger-Müller counters, greatly increasing the lifetime of the detector tube.

1949   In December, the uranium rush is created by the U.S. Atomic Energy Commission by placing an article in a mining journal offering to buy all the uranium that can be found in the continental United States.

1950   Using vacuum-tube technology and analog-to-digital conversion techniques, a 99-channel pulse-height analyzer is built for spectral measurements of radiation energy.

1952   The National Bureau of Standards, using newly formulated calculation methods, issues recommendations for lead-shielding thicknesses, published as document NBS-1003.

1953   On May 12, a TX-9 nuclear artillery shell, code-named Grable, is detonated aboveground in the Upshot-Knothole exercise in the Nevada desert. After the photon blast and the delayed shock wave have passed, hundreds of soldiers are ordered out of their protective trenches to walk directly toward the ground zero detonation point.

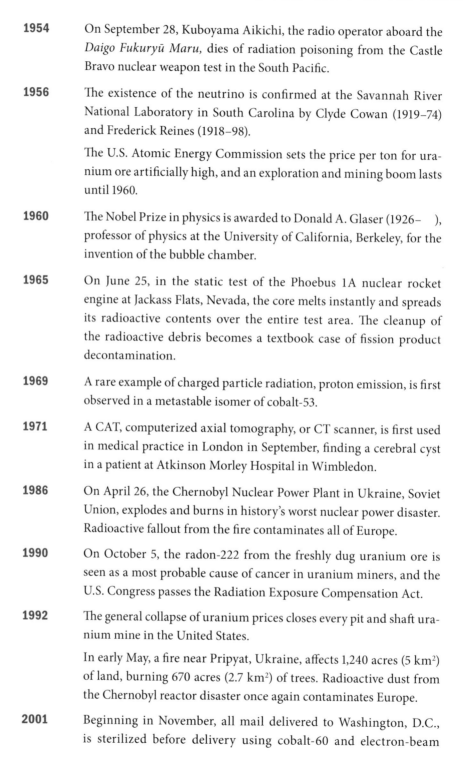

1954   On September 28, Kuboyama Aikichi, the radio operator aboard the
            Daigo Fukuryū Maru, dies of radiation poisoning from the Castle
            Bravo nuclear weapon test in the South Pacific.

1956   The existence of the neutrino is confirmed at the Savannah River
            National Laboratory in South Carolina by Clyde Cowan (1919–74)
            and Frederick Reines (1918–98).

            The U.S. Atomic Energy Commission sets the price per ton for ura-
            nium ore artificially high, and an exploration and mining boom lasts
            until 1960.

1960   The Nobel Prize in physics is awarded to Donald A. Glaser (1926–   ),
            professor of physics at the University of California, Berkeley, for the
            invention of the bubble chamber.

1965   On June 25, in the static test of the Phoebus 1A nuclear rocket
            engine at Jackass Flats, Nevada, the core melts instantly and spreads
            its radioactive contents over the entire test area. The cleanup of
            the radioactive debris becomes a textbook case of fission product
            decontamination.

1969   A rare example of charged particle radiation, proton emission, is first
            observed in a metastable isomer of cobalt-53.

1971   A CAT, computerized axial tomography, or CT scanner, is first used
            in medical practice in London in September, finding a cerebral cyst
            in a patient at Atkinson Morley Hospital in Wimbledon.

1986   On April 26, the Chernobyl Nuclear Power Plant in Ukraine, Soviet
            Union, explodes and burns in history's worst nuclear power disaster.
            Radioactive fallout from the fire contaminates all of Europe.

1990   On October 5, the radon-222 from the freshly dug uranium ore is
            seen as a most probable cause of cancer in uranium miners, and the
            U.S. Congress passes the Radiation Exposure Compensation Act.

1992   The general collapse of uranium prices closes every pit and shaft ura-
            nium mine in the United States.

            In early May, a fire near Pripyat, Ukraine, affects 1,240 acres (5 km²)
            of land, burning 670 acres (2.7 km²) of trees. Radioactive dust from
            the Chernobyl reactor disaster once again contaminates Europe.

2001   Beginning in November, all mail delivered to Washington, D.C.,
            is sterilized before delivery using cobalt-60 and electron-beam

accelerators. This extreme precaution is specifically used to kill anthrax bacteria that may be sent through the mail to political offices in the nation's capital.

2009   ATLAS, a multipurpose radiation detector assembly, is turned on at the Large Hadron Collider on the border of Switzerland and France.

2011   On March 11, a magnitude 9.0 earthquake and subsequent tsunami inundate the Fukushima-1 nuclear power plant on the northeastern coast of Japan. An unprecedented three power reactors melt down due to flooded backup generators, and the nearby farming community is contaminated with iodine-131 and cesium-137.

Glossary

alpha particle also alpha rays, a class of ionizing radiation composed of helium nuclei traveling at high speed. Alpha particles have a charge of +2 and are composed of two protons and two neutrons traveling stuck together. Alpha particles are highly energetic but have little ability to penetrate anything. Alpha particles are emitted from heavy nuclei undergoing severe decay.

atomic bomb or A-bomb, an antiquated term meaning a nuclear weapon using a prompt, fast fission reaction in U-235 or Pu-239 as the explosive agent. A better term is nuclear weapon or nuclear device.

background radiation equivalent time (BRET) a way of expressing medical X-ray exposure by referencing the amount of time spent under natural background radiation to absorb the same dose of radiation

barn A whimsical comment that part of a nuclide was big as a barn door led to this unit, defined as

$$1 \text{ barn (bn)} = 10^{-24} \text{ cm}^2.$$

becquerel (Bq) the SI unit of radiation measurement; one becquerel equals 2.7×10^{-11} curies, or one nuclear disintegration per second.

beta ray or beta particle, either an electron or a positron ejected from a decaying nucleus. If it is an electron, then a neutron has decayed into a proton. If it is a positron, then a proton has decayed into a neutron.

BF3 detector a special radiation detection tube filled with boron trifluoride gas. When hit by a thermal neutron, the boron activates and then immediately decays, causing a detectable radiation ionization in the tube. This technique is used to detect and count neutrons.

big bang a comprehensive explanation of the beginning of the universe, based on simplifications of the theory of general relativity by Albert Einstein (1879–1955). The big bang is thought to have occurred suddenly 13.7 billion years ago.

bremsstrahlung breaking radiation, or the electromagnetic radiation caused by the sudden negative acceleration of a high-speed electron as it crashes into a metal obstacle

chain reaction a series of chemical or nuclear reactions in which each reaction causes another reaction. A physical map of the reactions will show them connected, as if in a chain.

CNO cycle (carbon-nitrogen-oxygen) fusion reaction by which stars convert hydrogen to helium; also known as the Bethe-Weizsäcker cycle

cobalt-60 a radioactive isotope of cobalt; cobalt-60, or Co-60, emits a powerful gamma ray with a half-life of 5.27 years.

cosmic rays extremely high-energy radiation particles, usually protons, originating in outer space and constantly striking the Earth's atmosphere. Cosmic rays can have energies of more than 10^{20} electron volts, which is far higher than any particle energy that has been achieved by artificial means.

curie a unit of radioactivity, defined as

$$1\,Ci = 3.7 \times 10^{10}\text{ decays per second}$$

daughter product the result of the radioactive decay of a nucleus

decay chain the series of steps through which a radioactive isotope progresses, becoming different and successively lighter isotopes, as it decays toward a final, stable, nonradioactive isotope

delta ray secondary radiation similar to beta particles caused by free electrons colliding with laboratory apparatus, as discovered by J. J. Thomson (1856–1940) in 1905

deuterium heavy hydrogen. The deuterium is heavy because the nucleus contains both a neutron and a proton; it weighs twice what an ordinary hydrogen nucleus weighs.

dosage the integral or cumulative amount of radiation absorbed over a period of time

electromagnetic radiation a form of radiation brought into being by the alternating creation of a magnetic field and an electric field. The electric field creates a magnetic field, and the magnetic field creates an electric field. The two alternating processes move outward in a sphere, moving at the speed of light.

electrometer an instrument used to detect small electrical currents

electroscope an instrument used to detect electrical charge, usually by the principle of repulsion of two like-charged objects

element a pure chemical substance, consisting of one type of atom. An atom type is determined by the number of protons in its nucleus, and this configuration is unique to each of 118 elements known to exist.

enriched uranium uranium reactor fuel that has had the U-235 content improved. A typical power reactor uses fuel with the U-235 artificially increased to 8 percent.

extremely low frequency (ELF) the lowest frequency radiation in the electromagnetic spectrum. ELF waves vibrate at three to 3,000 hertz. Naturally occurring ELF waves are 7.8 hertz, with a wavelength of 24,819 miles (4×10^7 m).

fallout radioactive dust made in a nuclear weapon explosion by neutron activation of native materials, fission products from the explosion process, or unused radioactive material in the weapon, spread out over a large land area by atmospheric transport

fissile a descriptor for an element that will release energy and excess neutrons when fissioned

fission the splitting of a heavy nucleus into two lighter nuclei. Fission is caused by the absorption of a neutron in fissionable elements and can result in the release of excess energy.

fission products the lighter, always radioactive isotopes into which a fissile fuel breaks upon fission. Fission products are a wide range of isotopes, with half-lives from a few seconds to a few thousand years.

fluorescence an energy conversion event at the atomic level, causing a visible glow. Ultraviolet light, for example, will make fluorescent dye glow, as the energetic ultraviolet photon is down-converted to a visible photon, reemitted by the dye.

fusion the combining of two light nuclei into one heavier nucleus, resulting in a release of excess energy

gamma ray a high-energy electromagnetic wave, above X-rays on the electromagnetic energy spectrum, originating in the nucleus. A gamma ray is generated when the nucleus experiences a rearrangement of subnuclear particles.

Geiger counter or Geiger-Müller counter, is an electronic radiation detector used to measure the presence of gamma or beta rays. The Geiger counter makes use of the extreme amplifying properties of an avalanche effect in a gas-filled tube excited by a high voltage.

giga electron volts (GeV)   one billion electron volts

gray   the SI unit of energy absorbed due to radiation exposure; one gray equals 100 rads, or the absorption of one joule of energy by one kilogram of matter.

half-life   the time required for a radioactive sample to decrease its level of radioactivity by one half

high frequency (HF)   radio waves in the band from three to 30 megahertz. Also known as the shortwave band. The waves range from 100 to 10 meters long.

hormesis   an unproven, hypothetical process in which human exposure to low levels of radiation improves resistance to high levels of radiation

infrared   light not visible to the human eye, as it is below red on the spectrum. Infrared light vibrates at tens to hundreds of terahertz, at wavelengths from 100 micrometers to 750 nanometers.

International Atomic Energy Agency (IAEA)   an oversight organization monitoring nuclear activities; based in Vienna, Austria

ionization chamber   a metal can, filled with a gas, with an electrode in an insulated position in the center, used to detect ionizing radiation by its tendency to ionize the gas between the outer can and the inner electrode. Weak electrical conduction between the can and the electrode is induced by the ionization.

ionizing radiation   radiation of sufficient power to knock the top electron out of an atom upon collision. Examples of ionizing radiation are gamma rays, beta rays, and alpha rays.

isotope   a subspecies of an element, distinguished by the number of neutrons in the nucleus. All possible isotopes of hydrogen, for example, have zero, one, or two neutrons in the nucleus, all of which have only one proton.

joule   a unit of energy, defined as one newton of force acting to move an object through one meter of distance. It is equal to one watt-second.

kilo electron volt (KeV)   1,000 electron volts

light water   ordinary water, as is available from a municipal tap. Light water contains only traces of heavy water, or deuterium oxide.

low frequency (LF) radio waves in the electromagnetic spectrum from 30 to 300 kilohertz. Wavelengths range from 10 kilometers to one kilometer. Commercial radio in this band consists of amplitude modulated signals in Europe, Africa, and the Middle East.

mass number also atomic mass number or nucleon number, the total number of neutrons and protons in an atomic nucleus

medium wave (MW) an electromagnetic radiation band used for AM radio broadcasts, ranging from 530 to 1,700 kilohertz. The wavelength is about 300 meters.

metastable isomer a nuclide that is in a meta state. It is possible for an isotope to have two associated nuclides, or isomers. One is a metastable isomer, and the other is the result of metastable decay, or the stable isomer.

meta state the unstable state of a nucleus, causing it to undergo gamma decay without altering its mass number. The nucleus seeks a more stable configuration by rearranging its neutrons and protons, and the acceleration of one or more protons results in electromagnetic radiation in the form of a gamma ray.

million electron volts (MeV) unit of energy applied to subatomic or subnuclear particles in motion

N factor used to find dose of irradiated tissue, depending on part of body affected, the time and volume of the spread of radiation, or the species

neutrino an elementary particle of matter, of the lepton group, having zero electrical charge and an extremely small mass. Neutrinos have only the slightest tendency to interact with matter and are thus extremely difficult to detect.

neutron a fundamental particle of matter having no electrical charge and a mass slightly larger than that of a proton. Neutrons are components of the nucleus in an atom.

nucleosynthesis the building of heavy, complex elements in stars by nuclear fusion

nucleus the massive center of an atom, built of protons and in all but one case, neutrons. Hydrogen is the only atomic nucleus having no neutrons and only one proton.

nuclide a subspecies of an element. The element is defined by the number of protons in a nucleus. The nuclide, or the isotope, of an element is defined by the number of neutrons in the nucleus. Nuclide is a finer distinction, because one isotope can have more than one nuclide, due to metastable forms.

photon a particle of electromagnetic radiation. The photon is the particle representation of this radiation, which may be alternately represented by a ray.

pion (short for pi mesan, denoted with π) subatomic particles, in the group of mesons, in the statistical grouping of bosons. Designated π^0, π^+, and π^-, they are part of the debris caused by the collisions of cosmic rays with air molecules in the upper atmosphere.

Planck's constant a very small number used to express the sizes of quanta in quantum mechanics. The classic Planck's constant, named for the German physicist Max Planck (1858–1947), is about 6.6×10^{-34} joule-seconds.

plutonium element number 94 in the table of the elements; plutonium is a chemically poisonous metal and exceedingly rare in nature. Plutonium is commonly made by activating U-238, which decays to neptunium, which then decays to plutonium.

Pu-239 a fissile isotope of plutonium; one Pu-239 nucleus contains 94 protons and 145 neutrons.

Q relative biological effectiveness, or a factor used to adjust radiation measurements to reflect the dangerous effects of various radiation types on different biological systems

rad a unit of absorbed radiation dose; one rad is enough radiation to deposit 100 ergs of energy in one cubic centimeter of water.

radioactive decay the tendency of certain isotopes to undergo change in the nucleus. Any change in the nuclear structure causes radiation to be emitted from the nucleus. The time at which the change occurs is completely random and unpredictable, yet the rate at which a large sample of the particular isotope will decay is predictable and characteristic of the isotope.

radioactivity the emission of radiation, either by the willful manipulation of a nucleus or by spontaneous nuclear decay

radiography diagnostic examination of internal body structures by studying the X-ray shadows cast by bones and other radiation-blocking materials

radioisotope a subspecies of an element that is unstable and subject to nuclear decay

radiolysis   the decomposition of water into hydrogen and oxygen by the action of radioactivity

radionuclide   See radioisotope

radiopharmaceutical   a radioactive substance licensed for use as a medicine to treat or diagnose disease

radura   a symbol attached to food products, indicating that the item has been irradiated. The symbol is a green circle pierced with five radiating lines, enclosing the image of a green flower.

roentgen   a unit of radiation measurement; one roentgen of ionizing radiation is enough to completely ionize one cubic centimeter of dry air at standard temperature and pressure.

roentgen equivalent man (rem)   A unit of radiation dose adjusted for its effectiveness against human flesh. One rem equals 1.07185 roentgens.

self-shielding   the tendency for a solid mass of radioactive material to capture its own radiation internally and not allow it to escape into the surrounding space. Only the surface of a radioactive mass can actually emit radiation, with the bulk of the radiation converting to heat inside the mass.

sievert   the SI measurement of radiation dose. One sievert equals 100 rem.

supernova   the violent death of a star, resulting in the expulsion of more radiation per second than the output of an entire galaxy. Elements heavier than iron are made by fusion in a supernova explosion.

survey meter   a radiation meter used to monitor and detect dangerous conditions in a radiation work environment

Système international d'unités (SI)   the international metric system of measurements, nearly globally adopted

tera electron volts (TeV)   1 trillion electron volts

terahertz radiation (T-rays)   occurring in the electromagnetic spectrum between microwaves and infrared light in the frequency band from 300 gigahert to three terahertz. Wavelengths are between one millimeter and 100 micrometers. The terahertz band is an active subject of research and application.

thermal neutron   neutron slowed to the point where it is moving no faster than surrounding molecules; at room temperature, molecules and thermal neutrons move with an energy of 0.025 electron volts.

thorium-232   the naturally occurring isotope of thorium; it alpha decays into radium-228 with a 1.4×10^{10} year half-life.

tomography   a means of reconstructing an image of hidden internal structure of an object using nothing more than measurements of its density through a planar slice

tritium   the heaviest isotope of hydrogen, having two neutrons in the nucleus. Tritium is slightly radioactive. It emits a low-energy beta particle.

U-235   a fissile isotope of uranium, having 92 protons and 143 neutrons in its nucleus. U-235 occurs rarely in nature, making up only 0.7 percent of the uranium found in Earth's crust.

U-238   a nonfissile isotope of uranium, having 92 protons and 146 neutrons in its nucleus. U-238 makes up most of the uranium occurring in nature. It can decay indirectly into Pu-239 upon neutron capture.

ultrahigh frequency (UHF)   radio waves in the band from 300 megahertz to three gigahertz, with wavelengths from one meter down to 10 centimeters. Radio in this band is used for global positioning system (GPS) satellite-to-ground communications and WiFi Internet connections.

ultraviolet light   light that is beyond the frequency of violet light, and is therefore invisible to human eyes

uranium   element number 92 in the table of the elements; it is common in nature and can be found in traces in most of Earth's crust. It is mildly radioactive.

very high frequency (VHF)   radio waves in the band from 30 to 300 megahertz, at wavelengths from 10 to one meters. FM radio and television signals are carried on the VHF band.

X-ray   an electromagnetic wave above light in the electromagnetic energy spectrum. An X-ray originates in the electron cloud surrounding a nucleus in an atom or can be generated by rapidly accelerating a freely moving electron.

Ahmed, Syed Naeem. *Physics & Engineering of Radiation Detection.* San Diego, Calif.: Academic Press, 2007. A complete description of current methods of detecting radiation, on the graduate level.

Field Manual FM 3-3 (NAVFAC P-462), *NBC Contamination Avoidance.* Washington, D.C.: Headquarters Department of the Army, 1986. Written for the armed forces, this manual contains much practical information for detecting and avoiding accidental or intentional radioactive contamination. It includes nonograms for use in calculating the decay and total dose from radioactive fallout using a straightedge.

Field Manual FM 3-14 (MCRP 3-37.1A), *Nuclear, Biological, and Chemical Vulnerability Analysis.* Washington, D.C.: Headquarters Department of the Army, 1997. This military field manual contains an assessment of vulnerabilities to nuclear attack by nuclear explosives or radiological "dirty" bombs. It includes radiation exposure risk analysis and reduction measures.

Gibson, James Norris. *The History of the U.S. Nuclear Arsenal.* Greenwich, Conn.: Brompton Books, 1989. Every nuclear weapon ever developed and owned by the United States is described in this complete history. There was apparently an atomic bomb for every occasion.

Glasstone, Samuel. *The Effects of Nuclear Weapons.* Washington, D.C.: U.S. Government Printing Office, 1957. This classic remains the last word on the details of what an aboveground nuclear weapon explosion will do to living things and the physical plant in which we live.

Goldstein, Donald M., Katherine V. Dillon, and J. Michael Wenger. *Rain of Ruin: A Photographic History of Hiroshima and Nagasaki.* McLean, Va.: Brassey's, 1995. The bombings of Hiroshima and Nagasaki, Japan, in 1945 were partially scientific experiments to determine the effects of nuclear weapons. This is a report on those findings.

Hansen, Chuck. *U.S. Nuclear Weapons: The Secret History.* New York: Orion Books, 1988. Every fact, diagram, and photograph from the nuclear weapons development programs of the United States that was declassified by 1988 is in this book. Speculative items, such as the design of the L-11 "Little Boy" atomic bomb have since been proven inaccurate, but most technical details in this book are accurate and not available elsewhere.

RECOMMENDED WEB SITES

Further depth on many topics covered in this book has become available on the Web. Available are detailed Web sites for any government facility or agency, some giving access to archives and histories. The books listed above can be bought online, and there are even entire books concerning nuclear power issues, long out of print or published by the Government Printing Office, that are available. Although the Web is constantly expanding and changing and it would not be possible to cite a fraction of the important links to nuclear topics, the following list is a start.

Alternate Civil Defense Museum, or southernradiation/Paul's RADIC Showcase, is a unique listing of civil defense and military radiation detection equipment built specifically for the possibility of nuclear warfare. All civil defense radiation detectors built in the United States are included, but there are also complete compilations of equipment used in Canada, the United Kingdom, Russia, Switzerland, Germany, and the Netherlands. Instruction manuals and color photographs show both antique and currently used equipment. Available online. URL: http://civildefense museum.org/southrad/index.html. Accessed February 9, 2011.

American Nuclear Society offers a free, interactive radiation dose chart that will calculate your exposure to radiation based on living location, jet travels, medical X-rays, and other factors. This same dose calculator is also offered as an iPad application, available on iTunes. Available online. URL: http://www.new.ans.org/pi/resources/dosechart. Accessed February 9, 2011.

Civil Defense Museum, established in 2003, covers all aspects of the cold war years, when the federal Civil Defense Administration was charged with preparing every citizen in the United States for a nuclear exchange with the Soviet Union. Fallout shelters were built, food and medicine were stockpiled, and a sufficient number of radiation counters were built so that every family would have at least one. This site has everything from virtual tours of fallout shelters to complete listings of all the civil defense radiation counting equipment. Available online. URL: http://www.civildefensemuseum.com. Accessed February 9, 2011.

Goiânia Radiation Incident is covered extensively at this site, as written by Marco Antônio Sperb Leite and L. David Roper. Leite is a professor of physics at the Federal University of Goiás in Goiânia, Brazil, and Roper is a professor of physics at Virginia Polytechnic Institute and State University in Blacksburg, Virginia. This account of the cobalt-60 spill accident in 1986 covers a history of the Brazilian nuclear policy, radiation therapy devices for cancer treatment in Brazil, the future of the victims, and a complete chronology of the incident. Available online. URL: http://arts.bev.net/roperldavid/GRI.htm. Accessed February 9, 2011.

LANL Research Library is a vast collection of ebooks, databases, and news concerning all issues of nuclear science, nuclear technology, the history of nuclear topics, and current research and development. Students of nuclear science will find it an invaluable resource of information. The library search can find books, journals, patents, reports, videos, audiotapes, and recommended Web sites. Available online. URL: http://library.lanl.gov. Accessed February 9, 2011.

National Radiation Instrument Catalog was started in October 2001 as an effort to document every portable radiation counter that had been developed in the important period from 1920 to 1960. This Internet site was built in October 2007 to make this accumulated information available. The historical overview of radiation instruments is superb, and this depth of information and photographs is not available elsewhere. The catalog section of the site gives information on every known brand of radiation instrument, distributors, national laboratories, military organizations, and even atomic toys, kits, and novelties. The references section includes scientific and technical books concerning radiation detection, operating manuals for hundreds of radiation counters, patents, trade catalogs, military publications, and libraries. Available online. URL: http://national-radiation-instrument-catalog.com. Accessed February 9, 2011.

U.S. Department of Energy is a rich, complete site spanning the wide interests of this government agency, from energy science and technology to national security of energy sources. The Department of Energy owns, secures, and manages all the nuclear weapons in the military inventory. Although it concerns all sources of energy, nuclear power is a large component of the mission of the Department of Energy. This site is particularly accessible to students and educators. Available online. URL: http://www.doe.gov. Accessed February 9, 2011.

Index

Italic page numbers indicate illustrations; *m* indicates a map.